Greenhouse Gas Mitigation Assessment: A Guidebook

Environmental Science and Technology Library

VOLUME 6

Greenhouse Gas Mitigation Assessment: A Guidebook

Lead authors:

Jayant Sathaye

and

Stephen Meyers

Lawrence Berkeley National Laboratory,
International Energy Studies,
Berkeley, CA, U.S.A.

Chapter authors:

Barbara Allen-Diaz
Richard Cirillo
Michael Gibbs
Ed Hillsman
Willy Makundi
Jim Ohi

Contributing authors:

Gunther Conzelmann
David Corbus
Ashok Gadgil
Charles Heaps
Jon Koomey
Michael Lazarus
Jason Mark
Dennis Ojima
Marc Ross
Alan Sanstad
Fridtjof Unander
John Weyant

Springer-Science+Business Media, B.V.

A C.I.P. Catalogue record for this book is available from the Library of Congress.

ISBN 978-90-481-4634-5 ISBN 978-94-015-8466-1 (eBook)
DOI 10.1007/978-94-015-8466-1

Printed on acid-free paper

Originally published by Kluwer Academic Publishers in 1995.
Softcover reprint of the hardcover 1st edition 1995

TABLE OF CONTENTS

Chapter 15
REPORTING A MITIGATION ASSESSMENT

PREFACE

This book describes guidelines prepared by the U.S. Country Studies Program for the evaluation of options to mitigate greenhouse gas emissions. The U.S. Country Studies Program developed these guidelines in collaboration with Lawrence Berkeley National Laboratory to provide developing countries and countries with economies in transition with reference materials for national mitigation assessments. Over 50 countries participating in the program have used the guidelines, which have been refined to reflect their comments.

The guidelines delineate a step-wise methodology for evaluating greenhouse gas mitigation options for the energy and non-energy sectors and describe the applications of common analytical tools. The U.S. Country Studies Program uses these guidelines in conjunction with intensive training workshops and follow-up technical assistance during the lifetime of each country's study. The program uses similar reference materials to assist counties with their greenhouse gas emission inventories and evaluations of climate change vulnerability and adaptive responses.

These guidelines serve three purposes:

- to assist countries in making decisions about the scope and methodology for mitigation assessments;
- to provide countries with guidance and step-by-step instructions on each element of a mitigation assessment; and
- to help countries determine which analytical tools are best suited to their needs and describe procedures for applying these tools.

This book describes the application of the most common and readily available methods and analytical tools. Countries are encouraged, where appropriate, to use their own methods.

We are grateful to the country study teams for their advice on the development of the guidelines and applaud their commitment to improving global awareness and support for greenhouse gas mitigation.

Ron Benioff, Technical Director
U.S. Country Studies Program

ACKNOWLEDGMENTS

This guidebook is the result of a collaborative effort by many people. In addition to the Lead, Chapter, and Contributing Authors, the following people contributed material that was used in assembling the guidebook: James Krider, U.S. Country Studies Management Team, and Puran Mongia and Nandita Mongia, University of Delhi, India. A number of reviewers provided helpful comments on earlier drafts of the guidebook. In particular, Ron Benioff of the U.S. CSMT made helpful suggestions on many parts of the guidebook. Stephen Wiel of Lawrence Berkeley National Laboratory also contributed many useful comments.

The production of this guidebook was supported by the U.S. Country Studies Program. We acknowledge the leadership of Robert Dixon, Director of the Program, as well as the guidance of Chris Bordeaux, Manager of Mitigation Support for the Program. The guidebook would not have been possible without the efforts of Karen Olson of LBNL, who managed the production of the final manuscript, and Diana Morris of LBNL and Pam Condie of Compleat Small Business Services, who did the word processing. Lorraine Williams of LBNL produced a number of the figures and tables. Joe Wisniewski reviewed the final draft and coordinated publication.

AUTHOR AFFILIATIONS AND AREAS OF CONTRIBUTION

LEAD AUTHORS:

Jayant Sathaye and Stephen Meyers
Lawrence Berkeley National Laboratory
Berkeley, CA

CHAPTER AND CONTRIBUTING AUTHORS:

Barbara Allen-Diaz University of California, Berkeley Berkeley, CA	Rangelands and Grasslands
Richard Cirillo Gunther Conzelmann Argonne National Laboratory Argonne, IL	Conventional Energy Supply
David Corbus Jason Mark Jim Ohi National Renewable Energy Laboratory Golden, CO	Renewable Energy Supply
Ashok Gadgil Jon Koomey Willy Makundi Alan Sanstad Lawrence Berkeley National Laboratory Berkeley, CA	Agricultural Energy Use Residential/Commercial Energy Use Non-energy Sectors Energy Modeling
Michael Gibbs ICF Inc. Universal City, CA	Agriculture and Waste Management
Ed Hillsman Oak Ridge National Laboratory Oak Ridge, TN	Transportation Energy Use
Michael Lazarus Charles Heaps Tellus Institute Boston, MA	Energy Modeling

Dennis Ojima Colorado State University Fort Collins, CO	Rangelands and Grasslands
Marc Ross University of Michigan Ann Arbor, MI	Industrial Energy Use
Fridtjof Unander Institute for Energy Technology Kjeller, Norway	Energy Sector
John Weyant Stanford University Stanford, CA	Energy-Economy Models

NOTATION

The following is a list of the acronyms, initialisms, and abbreviations (including chemical symbols and units of measurements) used in this book.

ACRONYMS, INITIALISMS, AND ABBREVIATIONS

AFBC	atmospheric fluidized-bed combustion
AHP	Analytical Hierarchy Process
BRAC	Benefit of Reducing Atmospheric Carbon
BOD	biochemical oxygen demand
CCC	cost of conserved carbon
CCE	cost of conserved energy
CENR	Committee on Environment and Natural Resources
CGE	computable general equilibrium
CHP	combined heat and power
COMAP	COmprehensive Mitigation Assessment Process for forestry
COPATH	Carbon, Pasture, Agriculture, Total, Harvesting
CSMT	Country Studies Management Team
DOC	degradable organic carbon
DOE	U.S. Department of Energy
DOS	U.S. Department of State
EMS	energy management systems
ENPEP	ENergy and Power Evaluation Program
EPA	U.S. Environmental Protection Agency
EPIC	Erosion and Productivity Impact Calculator
FAO	Food and Agricultural Organization
GEF	Global Environment Facility
GDP	gross domestic product
GHGs	greenhouse gases

GWP	global warming potential(s)
HVAC	heating, ventilation, and air conditioning
IAEA	International Atomic Energy Agency
IGCC	integrated gasification combined cycle
IPCC	Intergovernmental Panel on Climate Change
LEAP	Long-Range Energy Alternative Planning
LIEF	Long-run Industrial Energy Forecasting model
LUCS	Land-use and Carbon Sequestration model
MARKAL-MACRO	Market Allocation Macro-economic model
MIMEC	Multisectoral, Intertemporal Model of Emissions Constraints
MSW	municipal solid waste
NGO	nongovernmental organization
NOAA	National Oceanic and Atmospheric Administration
O&M	operating and maintenance
OECD	Organization for Economic Cooperation and Development
ORNL	Oak Ridge National Laboratory
OTA	Office of Technology Assessment
PC	pulverized coal
PFBC	pressurized fluidized-bed combustion
PVC	polyvinyl chloride
RET	renewable energy technologies
SOC	soil organic carbon
SRF	short-rotation forestry
STAIR	Services, Transport, Agriculture, Industry, and Residential energy model
T&D	transmission and distribution
UEC	unit energy consumption
UNEP	United Nations Environment Programme

USAID	U.S. Agency for International Development
USDA	U.S. Department of Agriculture
VA	value added
VS	volatile solids
WRDC	World Radiation Data Center

CHEMICAL SYMBOLS AND ABBREVIATIONS

C	carbon
Ceq	carbon equivalent
CO_2	carbon dioxide
CO	carbon monoxide
CFC	chlorofluorocarbons
CH_4	methane
H_2	hydrogen
HCFC	halochlorofluorocarbons
NMHC	non-methane hydrocarbons
NO_x	nitrogen oxides
N_2	nitrogen
N_2O	nitrous oxide
O_2	oxygen
O_3	ozone
SO_x	sulfur oxides

UNITS OF MEASUREMENT

Btu	British thermal unit
^{o}C	degree(s) Celsius
d	day(s)
g	gram(s)
GJ	gigajoule
ha	hectare(s)
km	kilometer(s)
kW	kilowatt
kWh	kilowatt-hour
L	liters
m	meter(s)
m^3	cubic meter(s)
MWe	megawatt electric
Pg	petagram(s)
RPM	rotations per minute
tC	tonnes of carbon
Tg	teragrams
TOE	tonnes of oil equivalent
TWh	terawatt-hours
yr	year

CHAPTER 1
INTRODUCTION

1.1 INTRODUCTION

The Framework Convention on Climate Change commits Parties to the Convention to develop national programs and measures to respond to climate change. One of the key responses that countries can make is to adopt measures that can reduce atmospheric accumulation of greenhouse gases (GHG) and thereby delay the predicted impact of GHG on global climate. Such measures may either reduce GHG emissions (abatement) or increase terrestrial storage of carbon (sequestration). Because these measures can moderate GHG, they are termed "mitigation" options.[1]

A mitigation assessment involves a national-level analysis of the potential costs and impacts of various technologies and practices that have the capacity to mitigate climate change. Two key goals of an assessment are (1): to provide policy makers with an evaluation of those technologies and practices that can both mitigate climate change and also contribute to national development objectives, and (2) to identify policies and programs that could enhance their adoption. An initial mitigation assessment should be followed by more detailed evaluation of specific policies, programs, or projects designed to encourage implementation of selected technologies and practices.

The purpose of this book is to provide guidance in designing and conducting national mitigation assessments. Its main objectives are to assist in:

- deciding on the scope of the mitigation assessment and the methods to be applied;
- identifying, screening, and characterizing technologies and practices that have potential to mitigate climate change and also meet national development objectives;
- analyzing the potential costs and impacts of various technical or policy measures on net GHG emissions, socio-economic conditions, and national environmental quality; and
- identifying policies and programs that have potential to encourage adoption of attractive mitigation technologies and practices.

The book focuses on two broad topical areas: the energy sector (demand and supply) and non-energy sectors (forestry, agriculture, rangelands, and waste management). For each sector, it provides guidance on the evaluation of individual mitigation options and on the use of models for sectoral and integrated analysis of options. It describes data requirements, potential data sources, and approaches for developing data.

This book is designed especially for developing countries and countries with economics in transition. However, the technique can be applied in any country. It is intended for use along with training workshops.

This book is intended to complement the material on methods for mitigation assessment developed by the IPCC. It presents the challenges faced by developing countries and economies in transition in using mitigation methods, links the choice of specific methods to the decision-making process, identifies key questions in structuring the assessment process, and discusses the types of methods available, emphasizing the issues associated with their use and their application to specific sectors.

---

[1] The type of mitigation option referred to in this context is different from a measure that might be taken to mitigate or moderate the impacts of climate change (for example, on agriculture or coastal settlements). The latter type of measure is generally termed an adaptation option.

The material in this book complements the IPCC work by providing more detailed and practical guidance on the use of tools and methods which can be applied for analysis of mitigation options. Whereas the IPCC efforts lay out a broad range of information applicable to all levels of decision-making for assessing mitigation options, including the setting of national goals and the design of programs and projects, this book is designed to serve as a basic reference for the analysts who are actually conducting a national mitigation assessment.

1.2 FORMAT OF THIS BOOK

Chapter 2 presents an overview of the mitigation assessment process and explains the basic concepts and methods used in the evaluation of options. It provides guidance on setting basic parameters for use in a mitigation assessment.

Chapters 3-9 and Chapters 10-14 focus on analysis of mitigation options in the energy and non-energy sectors, respectively. Each chapter contains guidance on conducting analysis of various kinds of mitigation options, and on the development of future scenarios.

Chapter 15 provides guidance on reporting results from the energy and non-energy analyses.

CHAPTER 2
BASIC METHODS AND CONCEPTS FOR MITIGATION ASSESSMENT

2.1 PREPARING FOR A MITIGATION ASSESSMENT

Key factors that need to be considered in preparing a mitigation assessment are discussed below:

- **Defining the Time Frame of the Assessment.** A mitigation assessment typically focuses on long-run opportunities for reducing GHG emissions or enhancing carbon sinks since it takes time for changes that could affect GHG emissions in a significant way to be adopted. In addition to a long-run focus, a country might also evaluate near-term policy and program options.

- **Defining the Scope of the Assessment.** A mitigation assessment may include a variety of areas. These include energy demand and supply, forestry, agriculture, rangelands, and waste management. Ideally, an assessment should include analysis of the impact of mitigation options (particularly in the energy sector) on the macro-economy. Countries should structure their assessment to address the topics of most importance, taking into consideration the resources available to perform the study. A mitigation assessment should include some consideration of policies and programs that can encourage adoption of mitigation technologies and practices. More detailed evaluation of particular policy/program options could follow an initial assessment.

  Defining Primary Users of the Assessment. Countries should design their assessments to satisfy the needs of the various possible users or stakeholders. The primary users of the assessment are likely to be policy decision-makers who are responsible for evaluating and designing mitigation policies. The country's scientific community is likely to benefit from participation in the assessment process, and also from the compilation of data and access to new models, which will also be useful for other types of analysis. Other potential in-country users include the NGO community. In addition, the output of each assessment may be shared with interested groups in other countries, and regional and international organizations.

- **Defining Results that Meet the Users' Needs.** Defining the type of output desired from an assessment will help in selecting the areas where efforts should be focused. The output of a typical assessment will consist of economic, GHG, local environmental, and social impacts of mitigation options. It could also include a discussion of barriers to implementation of mitigation options and description of policies to overcome them. The importance to be placed on characterizing each type of output should be determined in consultation with the potential users of the assessment. Each team should also consider how best to present the results (e.g., journal articles, reports, briefings, workshops).

- **Selecting Approaches that are Consistent with Data Availability and Expertise of Researchers.** The sophistication of the analytical methods that will be used in the assessment will depend, in part, on the desired level of output detail. If approximate estimates of scenarios are sufficient, then detailed costing models may not be necessary.

The level of detail will determine the type and extent of data that need to be collected for the assessment. Often it may be difficult to collect new data through surveys or detailed technology characterizations. However, every attempt should be made to collate existing data, which often tend to lie unused. Alternatively, data may be collected through informal surveys, e.g., of technology manufacturers and energy users.

2.2 THE STRUCTURE OF A MITIGATION ASSESSMENT

The structure of a mitigation assessment will vary depending upon its goals and scope. The type of institutions involved in the study may also affect the structure. Studies in specific sectors such as energy or forestry can be conducted somewhat independently, yet it is important for all members of the study team to use common assumptions regarding basic parameters such as population and economic growth.

Although most of the work will be independent, interaction among the sector specialists can be beneficial. For example, the forestry specialists can provide the energy specialists with information on biomass resources that may be available for energy consumption in the future while the energy specialists can give the forestry experts information on the future demand for fuelwood. In addition, it is necessary for the analysts in sectors such as forestry, agriculture, and rangelands to work together to develop scenarios of how available lands may be utilized in the future.

The nature of a mitigation assessment will vary among sectors, but it is possible to describe a basic structure that illustrates the key components of an assessment and how they relate to each other (Figure 2-1). The first step is to assemble data for the base year on the activities and technologies/practices that are associated with GHG emissions or carbon storage. The type of data and the framework in which they are organized will obviously vary among sectors. In each case, however, the assembly of base-year data should draw on and be consistent with the GHG emissions inventory.

Once the base year has been documented in some detail, the remainder of the assessment involves an evaluation of what might or could occur in the future. The development of scenarios of the future requires data on the activities that result in GHG emissions or shape opportunities for carbon storage. The types of data include production of key industrial products, the number of urban and rural households in the residential sector, the number of vehicles in transportation, and demand for land and forest products. Development of scenarios requires a projection of the future levels of each kind of activity. Such projections in turn draw on assumptions made about growth in population, GDP, and other macro variables.

For each type of activity or resource demand, there are generally a number of technologies or practices that can be employed, each having different implications for resource use and GHG emissions or carbon storage. Because the range of technologies and practices that could reduce GHG emissions or sequester carbon is large, it is helpful to conduct a screening of potential technology options to select those that will receive further analysis. Examples of criteria that may be used for screening are presented in Section 2.4.

Once options have been selected for inclusion in the assessment, it is necessary to characterize technologies and practices with respect to their costs and other features (e.g., performance, lifetime, environmental characteristics, labor and infrastructure requirements). The technologies or practices may include those that are already available or in use, as well as those that are expected to be available in the future.

The information on future activity levels and potential technologies/practices is used to define scenarios that describe future resource use or production under a certain set of conditions. Scenarios should also take into account current and likely policies that are of importance in each sector. These include policies affecting pricing of energy and other resources, land use, transport infrastructure, as well as general economic policies that influence diffusion of new technologies (such as policies affecting domestic competition and foreign investment). The methods used to construct scenarios range from simple approaches that require the analyst to make numerous judgements and "mechanically" assemble a scenario, to complex models that take the input data and select specific technologies and their penetration rates based on various criteria (e.g., optimization of system costs).

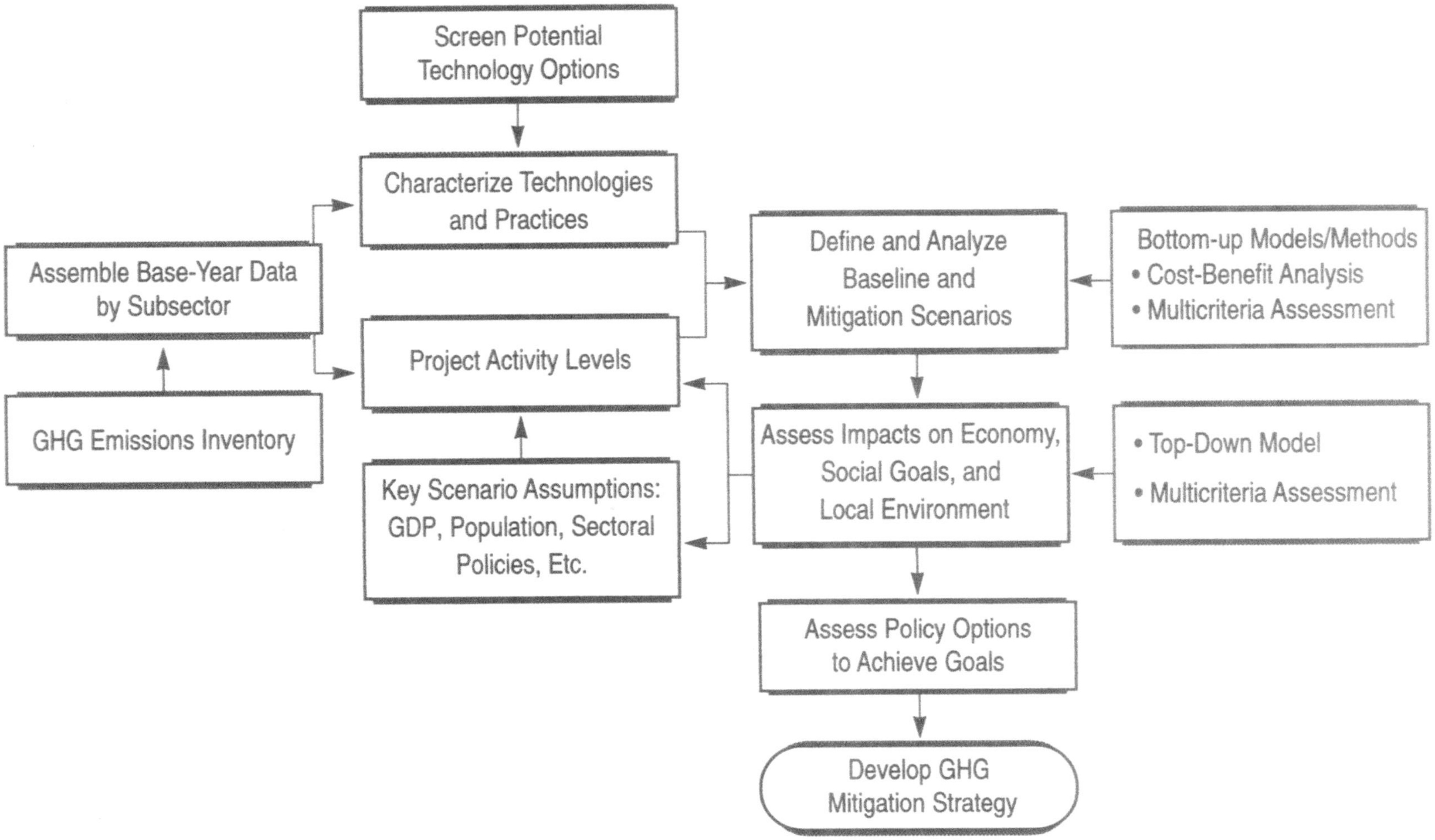

Figure 2–1. Structure of a Sectoral Mitigation Assessment

A mitigation assessment should include at least two scenarios for each sector considered. A "baseline" or "reference" scenario is a description of a plausible future in which no specific policy actions are taken to encourage actions that reduce GHG emissions or enhance carbon sinks. A "mitigation scenario" describes a future that is essentially similar to that in the baseline scenario with respect to overall economic and social trends, except that it assumes that policies or programs are implemented that encourage adoption of measures that will reduce GHG emissions or enhance carbon sinks (see Section 2.6 for further discussion of scenarios).

In analyzing the merits of mitigation options, standard techniques of benefit-cost analysis may be applied, with some modification (see Section 2.7). However, some of the impacts of a mitigation option may be difficult to express in monetary terms or even to quantify. Thus, cost-benefit analysis should be supplemented with quantitative and qualitative assessment of other criteria such as complementary environmental effects (e.g., reduction in local air pollution), secondary economic effects (e.g., employment creation), and social and political considerations (e.g., the impact on societal equity).

The combination of cost-benefit analysis and assessment of other criteria can be used to compare or rank mitigation options, which can support the definition of mitigation scenarios. A mitigation scenario may reflect only the technical potential of various options to reduce GHG emissions or to store carbon; or the analyst can estimate that part of the technical potential that may be achievable. Estimating the magnitude of GHG emission reduction or carbon storage that may be achievable within a given time frame requires identification and assessment of policies or programs that could be used to encourage adoption of mitigation options in each sector. A detailed evaluation of policies or programs may go beyond the scope of an initial mitigation assessment, however.

The scenarios provide estimates of the economic and other impacts within the sector being studied (e.g., investment requirements). It is important to also assess the impact of sectoral mitigation options (or of cross-sectoral options such as a carbon tax) on the overall economy, social goals, and the local environment. Various models may be used to analyze the interaction between the energy sector and the economy. Similar models have not been used to assess the impact on the economy of mitigation options in non-energy sectors, but models used for general analysis of agriculture and forestry could be modified for analysis of macroeconomic impacts of mitigation options in those sectors. In addition, methods that facilitate combined quantitative and qualitative assessment of multiple criteria may be used.

Once the full range of impacts of various mitigation options have been evaluated, the next step involves an assessment of policy and program options to encourage their adoption. This assessment may go beyond the rough assessment used to define mitigation scenarios. It might combine quantitative analysis with workshops that facilitate interaction between the analysts and relevant policy-makers, and other interested parties in a particular country. The goal of such interaction could be the development of a national GHG mitigation strategy for the sector.

The structure described above generally applies for the energy sector and the non-energy sectors. However, the analysis of the energy sector is complicated by the fact that it requires an integration of energy demand and supply in order to estimate the GHG impacts of mitigation options in the energy demand sectors, and to compare demand-side and supply-side options. It also tends to be larger in scope, since the use of energy pervades the entire economy.

2.3 TIME HORIZON IN MITIGATION ASSESSMENT[1]

The time horizon of the analysis plays a critical role in planning a mitigation assessment and selecting methods. For discussion purposes, it is useful to define time periods somewhat arbitrarily as near-term (1-5 years), mid-term (15 years), and long-term (15-50+ years).

A classic definition of the near term is that period of time during which the capital stock is fixed, while the technology mix and allocation of input factors across producing sectors is variable in the long term. Since the economy's structure and productive capacity are fixed in the short run, price changes are generated by fluctuations in variables reflecting seasonal cycles or transient impacts such as severe weather, oil spills, geopolitical conflicts, labor strikes, and so forth.

Extending the forecast horizon shifts the emphasis towards market clearing and relatively stable trends in variables. While the mid-term time frame extends far enough for innovations to occur in some sectors, the general characteristics of available technology, the capital stock of energy-consuming equipment, and demographic patterns can be anticipated with reasonable confidence. Although new technologies are continually emerging and their eventual market penetration is uncertain, gradual capital stock turnover implies that most of the capital equipment that will be used during the mid-term period is already in place, available on the market, or on the verge of commercialization.

Long-term modeling focuses on paths of key variables such as demographic patterns, cumulative impacts such as the depletion of nonrenewable resources, and changes in structural relationships between exogenous and endogenous variables due to capital stock turnover, the penetration of new technologies, emergence of alternative energy sources, and interindustry shifts in the composition of demand.

In the short term, only minor changes to the energy system can be anticipated to take place. Even in a mid-term perspective, substantial deviations from the existing planned energy system are limited due to long lag times for construction of supply facilities, and similar considerations apply for technology innovations in other sectors. Hence, the usefulness of advanced integrated models is less if one is addressing short-term impacts only. In this case, careful study of recent trends and evaluation of existing plans (expansion of the supply side is usually planned for the next decade) and available retrofit options, together with use of simple sector-specific models, is sufficient.

However, in GHG mitigation the real challenge is to be able to sustain reduced emission levels (or lower them further) in a long-term perspective. The goal of identifying opportunities for GHG mitigation that can have a significant impact in the future calls for long-term modeling of the energy and non-energy sectors. In the presentation of models in this book, we concentrate on models suited for addressing long-term development of the energy and non-energy sectors.

2.4 APPROACHES FOR ANALYSIS OF MITIGATION OPTIONS

Developing a national mitigation strategy requires identification and analysis of different actions that government could take to encourage adoption of technologies and practices that reduce GHG emissions or enhance carbon sinks. Based on the analysis, policy-makers can then decide which options not only satisfy specific policy objectives but are also within institutional, political, and budget constraints. Typically, the analytic process will follow a series of steps, each of which produces information for decision-makers. The manner in which these steps are performed will reflect each country's resources, objectives, and decision-making process.

---

[1] This section draws on Kydes *et al.* (1995).

Two general approaches have been used for mitigation assessment. The top-down method assumes a macroeconomic perspective wherein mitigation costs are defined in terms of losses in economic output, income, or GDP. A key assumption underlying many top-down analyses is that the baseline scenario represents the economy in equilibrium, with all factors of production employed efficiently given prevailing prices. The bottom-up approach focuses on individual processes such as end-use energy consumption, production of specific crops, and specific forest management schemes. For each relevant process, the approach attempts to estimate the costs associated with changes that result in GHG emission reductions or other impacts.

The fundamental difference between the two approaches is in the perspective taken by each on consumer behavior. The top-down approach assumes that consumers always act to maximize their utility or profit. According to this approach, if energy efficiency is less than it could be, it is because consumers see no economic gain in becoming more efficient. In contrast, the bottom-up approach assumes that various market barriers prevent consumers from taking actions that otherwise would be in their or the national economic self-interest. These market barriers could include lack of information about energy efficiency opportunities, lack of access to capital to finance the efficiency investment, and separation of responsibilities for making capital investments and paying operating costs (Lohani and Azimi, 1992).

In general, an assessment carried out using the bottom-up approach will very likely show significantly lower costs for meeting a given mitigation objective than will one using a top-down approach. To some extent, the differences may lie in a failure of bottom-up studies to accurately account for all costs associated with implementing specific actions. Top-down methods, on the other hand, can fail to account realistically for consumer and producer behavior by relying too heavily on aggregate data (Krause *et al.*, 1993). In addition, some top-down methods sacrifice sectoral and technology detail in return for being able to solve for general equilibrium resource allocations. Finally, top-down methods often ignore the fact that economies depart significantly from the stylized equilibria represented by the methods (Boero *et al.*, 1991). Each approach, however, captures costs or details on technologies, consumer behavior, or impacts that the other does not. Consequently, a comprehensive assessment should combine elements of each approach to ensure that most relevant costs and impacts are accounted for.

2.4.1 Tools for Mitigation Assessment

A variety of models and methods can assist the analysis of mitigation options. The primary ones described in this book are listed in Table 2-1. These range from bottom-up accounting models for the energy or forestry sector to top-down models of the whole economy in which energy or forestry are but one sector. For the energy sector, the models include accounting frameworks (LEAP and STAIR), optimization models (MARKAL and ETO), and an iterative equilibrium model (ENPEP). Each of these models may be used for integrated assessment of energy demand and supply although the approach and method varies among them. In addition to these quantitative models, a decision framework process (AHP) may be used for combined quantitative and qualitative evaluation of alternative technology options. These models are described in Chapter 3. Considerations for selecting among the different energy sector models are also presented in Chapter 3.

In industrialized countries, several top-down models have been developed for mitigation assessment, primarily to analyze the impact of carbon taxes. Similar models have not been widely applied to analyze changes in taxes, investment, or energy flows of developing and transition countries. Two top-down models that have been developed primarily for the developing countries are the MIMEC and LBL-CGE models. In addition, a recently developed hybrid model (MARKAL-MACRO) merges the bottom-up and top-down approaches.

In the non-energy sectors, bottom-up analytical methods have largely focused on the estimation of carbon and other GHG flows. The COPATH model has been used for carbon accounting and scenarios in the forestry sector, while COMAP has been developed for estimating the impacts of mitigation options in the forestry sector. EPIC and CENTURY are plant/soil simulation models which may be used to simulate carbon cycling dynamics in agricultural and rangeland ecosystems. For assessment of methane mitigation options in agriculture and waste management, simple spreadsheet models are available.

Top-down models have not been used in the non-energy sector for the assessment of GHG-related impacts in the developing and transition countries. Countries where agriculture or forest products form a significant share of the monetized economy have models or methods that can analyze the GDP impact of changes in domestic or international prices of these products. Many countries, for example, have used computable general equilibrium (CGE) models for analyzing the impacts on GDP, income distribution, and rural employment. These models may be modified for the purpose of a top-down analysis of the impacts of GHG mitigation options. The LBL-CGE model could also be modified for this purpose.

Table 2-1.
Examples of Analytical Tools Available for Mitigation Assessments

Topic	Analytical Tools
Energy Sector	
Accounting Models	LEAP, STAIR
Optimization Models	MARKAL, ETO
Iterative Equilibrium Model	ENPEP
Decision Analysis Framework	Analytical Hierarchy Process (AHP)
Non-Energy Sectors	
Forestry	COPATH, COMAP
Agriculture	EPIC, CENTURY
Rangelands	CENTURY
Waste Management	Landfill Gas Model
Energy-Economy Interaction	LBL-CGE, MARKAL-MACRO

2.5 SCREENING MITIGATION OPTIONS

The nature and importance of the screening process varies depending on the modeling approach. For each sector being studied, one approach is to develop a list of mitigation options that may be of interest. Various criteria, such as those listed in Table 2-2, are important for both screening and in-depth analysis of mitigation options. At the screening stage, one makes a rough assessment of the potential attractiveness of options, while the goal of the analysis is to quantify or carefully identify various impacts.

A useful approach for screening options is to prepare a matrix as shown in Table 2-2 for each sector. The matrix provides a qualitative indication of the attractiveness of each option by ranking it high, medium, or low, as judged according to each criterion. This matrix should be completed prior to conducting sectoral analyses in order to identify the options to be evaluated in depth.

Screening out the non-promising options requires careful judgement. An obvious reason for screening out options is if its wide-scale application is not viable. For example, location of options in environmentally or otherwise sensitive areas may rule them out for political reasons. The relationship between a mitigation option and development goals is important to consider. In addition, there may be options, such as reducing traffic congestion, which may be difficult to analyze since quantifying the impact on GHG emissions may be difficult to do. However, if the option is important for non-GHG reasons (e.g., as a measure to reduce urban air pollution), then simple assumptions may be made to roughly estimate its GHG impact.

Screening of options may require consideration of likely future conditions. For example, electricity-saving options may have a very small impact on GHG emissions if much of the electricity is hydro-generated. However, if the mix of generation is likely to shift toward more thermal generation, then electricity-efficiency options could become important.

2.6 DEFINING SCENARIOS

What will happen in the future cannot be predicted, but it is possible to develop scenarios of the future that reflect the consequences of different, but plausible, economic and technological conditions. For the purpose of mitigation analysis, at least two different scenarios are necessary. One scenario should reflect a baseline case, while the other should reflect the impact of mitigation options.

- **Baseline Scenario**. As mentioned earlier, a baseline scenario should represent a future in which there are no policies or programs designed to encourage or require actions that reduce GHG emissions or enhance carbon sinks. Defining a reasonable baseline scenario is a critical element in a mitigation assessment since the incremental costs and benefits of mitigation options will depend on the definition of the baseline scenario.

  A baseline scenario should not simply extrapolate from recent and current trends but rather incorporate a judgement of the likely evolution of resource-consuming and producing activities and technologies. This type of scenario is sometimes called "business-as-usual." Such a scenario would include some degree of adoption of technologies or practices that improve the efficiency of resource use and thereby reduce GHG emissions. In transition countries in particular, a baseline scenario will be quite different from historical trends.

  In both developing and transitional countries, where considerable economic and social change is expected over a period of the next several decades, it can be quite difficult to select a single image of the future as more likely than another. A study team might choose to define more than one baseline scenario. For example, alternate scenarios could reflect low, medium, and high economic growth. Obviously, there is a trade-off between keeping the assessment manageable and defining numerous baseline scenarios.

  If sectors of the economy contain structural inefficiencies (such as subsidized prices), it is important to be able to separate the costs and benefits of removing these inefficiencies from the impacts directly attributable to GHG mitigation measures. This ability to distinguish between general structural impacts and the impacts of mitigation policies is important regardless of whether an explicit policy of structural reform is assumed in the baseline scenarios.

Table 2-2. Criteria for Screening of Mitigation Options

Criteria	Mitigation Option 1	Mitigation Option 2	Mitigation Option n
Potential for large impact on CO_2 or other GHGs	High	Low	Medium
Direct cost/benefit ratio of the option	Low	High	High
Indirect economic impacts			
- Increase in domestic employment	Medium	Low	Low
- Decrease in import payments	Low	Medium	Uncertain
Consistency with national environmental goals			
- Reducing emissions of air pollutants	Low	High	Medium
- Effectiveness in limiting other environmental impacts	Medium	Low	Low
Potential ease of implementation	Low	Medium	High
Long-term sustainability of option	High	Uncertain	Medium
Consistency with national development goals	High	Low	Medium
Data availability for evaluation			
- Technology characterization	Low	Uncertain	High
- Costs of implementation programs	High	Low	Uncertain
Other sector-specific criteria	Low	High	Uncertain

Note: Numerical rankings may also be used.

- **Mitigation Scenario.** A mitigation scenario reflects a future in which climate-change mitigation is a primary motivation for adoption of technologies and practices that reduce GHG emissions or enhance carbon sinks. It may reflect only the technical potential for reducing GHG emissions or storing carbon, or it may incorporate estimates of what is achievable considering the many factors (institutional, cultural, legal, etc.) that may limit the implementability of the technically available options. Ideally, both the technical and the achievable potential should be reported.

  A study team could define and develop several mitigation scenarios. For example, alternate mitigation scenarios could reflect different degrees of emissions reduction or carbon storage relative to the baseline (e.g., 10%, 20%, 30%). A study team might also want to define mitigation scenarios that highlight particular types of technologies (e.g., renewable energy technologies).

  For both the energy and non-energy sectors, scenarios developed using a bottom-up approach take into consideration end-users' needs for energy, forest products, and land. By explicitly taking these needs into account, end-use scenarios are less likely to over- or understate final demand for products.

2.6.1 Setting Basic Parameters

In constructing scenarios, certain underlying parameters must be specified and treated consistently. Assumptions should be consistent with those used in GHG inventories and vulnerability and adaptation studies.

- **Selecting the Time Frame for the Assessment.** The time frame for GHG scenarios is often quite long, extending from 50 to 100 years. For mitigation options analysis, however, it is usually better to consider shorter time frames since the projection of macro-economic variables and the characterization of technologies beyond 20-30 years become quite uncertain. Analysts have generally used the period up to 2020 or 2030 as a relevant time frame to analyze the economics of mitigation options. For the forestry sector, long-rotation tree plantations may need to be evaluated over a longer time frame. Projections of emissions in the near-term (e.g., 2000) may also be helpful in evaluating policy options. Analysis may be conducted for either a single end year or several forecast years. In a dynamic framework, each consecutive time period is linked to the other over the entire time horizon.

- **Socio-Economic Variables.** Projections of socio-economic variables such as economic and population growth rates, land-use patterns, economic structure, and urban-rural population proportions may be obtained from national planning ministries in each country. Most countries have multi-year plans that show both economic structure and population growth assumptions. If these are not readily available, World Bank or UN projections of population and economic growth may be used instead. Economic growth projections are usually for a relatively short time period (5-10 years), and these should be extrapolated as realistically as possible for subsequent periods.

- **Land-use and Natural Resource Considerations.** Changes in land-use patterns will have an important bearing on GHG emissions from forestry, agriculture, and drylands, and they will also affect the vulnerability of the country to climate change. It is therefore important for both mitigation and vulnerability assessment staff to assess the current patterns of land use and their evolution over time. Changing the evolution of land-use patterns requires strong government policies and programs. Consideration should be made as to whether these types of policies are likely to occur as a mitigation option in each country. If only technical options are to be evaluated, then each "likely trends" scenario of land-use change should be used to evaluate both baseline and mitigation scenarios. If strong policies to modify evolving land-use patterns are plausible, then the mitigation scenario should consider a land-use pattern different from the one used for the baseline.

2.7 COST-BENEFIT ANALYSIS OF MITIGATION OPTIONS

A key objective of a mitigation assessment is to identify those options that maximize economic benefits or minimize the economic costs of restraining GHG emissions growth. Cost-benefit analysis has traditionally been used for project evaluation, but it has also been applied to mitigation assessment to estimate and compare relevant costs and benefits in a consistent and comprehensive manner. Cost-benefit analysis suggests that mitigation options that produce the greatest net benefit be selected among competing options. Strict application of this rule to the evaluation of mitigation options is not possible, however, since the benefits of mitigation options with regard to climate change cannot be monetized with any certainty at this point in time and are likely to vary among regions.

The monetizable portion of the costs and non-GHG benefits may be stated in money terms. Since the carbon GHG benefits cannot be easily monetized, the benefits may be stated simply in terms of either tonnes of carbon abatement or storage, or for non-carbon GHGs, in terms of carbon-equivalent.

Cost-benefit analysis should be supplemented by assessment of non-monetizable costs and benefits other than GHG abatement. These might include reduced emissions of other pollutants or an improvement in biodiversity. These costs and benefits should be quantified or at least described so that decision-makers can take them into account. Similarly, the impacts of an option on different societal groups may also need to be considered.

- **Discount Rate**. Cost-benefit analysis typically expresses costs and benefits that occur over a period of time in terms of their present value, which is calculated using a discount rate. The discount rate reflects the return on foregone present consumption that is sacrificed to secure future consumption. Since foregone present consumption is invested to secure future consumption, analysts often use a discount rate equal to the after-tax real rate of interest or return on capital investment.

  Much has been written about the estimation of discount rates for projects with long-term consequences (Lind *et al.*, 1982). For economic analysis of projects in the developing countries, real discount rates between 8 and 12% are commonly used by the World Bank. Each country should select an appropriate discount rate for evaluating the present value of monetary costs and benefits of mitigation options. A study team may wish to conduct sensitivity analyses at a higher and lower rate around the base rate. To evaluate options from the perspective of particular groups (such as households or farmers), analysts should use discount rates that are commonly used in their country for these groups.

  In order to assess an option's cost-effectiveness, the discounted costs and benefits are related to its GHG savings or carbon storage. Should the avoided GHG emissions or carbon storage be discounted at the same rate as costs? We suggest that these GHG flows not be discounted. By not discounting them, one assumes that the future economic damage caused by GHG emissions increases at the real rate of discount, which is not unreasonable considering that the potential damage that atmospheric GHG concentrations might cause in the future is largely unknown. (For a discussion of discounting monetary versus GHG flows, see Sathaye *et al.*, 1993.)

2.7.1 Cost Curves of GHG Mitigation

A GHG-reduction cost curve relates the quantity of GHG which can be reduced by mitigation options to the cost per unit GHG reduction. Correspondingly, a cost curve for stored carbon relates the quantity of stored, or sequestered, carbon to the cost per unit carbon stored. Cost curves for GHG reduction and for carbon storage can be combined to express the relationship between total amount of "avoided" GHG and the cost per unit GHG avoided.

Two distinct forms of cost curves appear in the literature: (a) discrete step curves and (b) continuous curves. These two forms are derived differently and should be interpreted in different ways. Within each of the two forms there are also different methods of construction, different meanings, and different interpretations. Schematic versions of the two forms are shown in Figures 2-2 and 2-3. The blocks in Figure 2-2 correspond to individual mitigation options or "baskets" of options, with widths representing the potential GHG reduction (or carbon stored) and heights representing the cost per unit GHG reduction. The points on the continuous curve (Figure 2-3) represent the increase in total system costs for a given scenario. The costs and the emission reductions represented by each point normally result from the contribution of a mix of several options, as opposed to the discrete step curve in which each option is analyzed separately. Continuous curves can be used to represent the aggregate output from models of the costs of reducing emissions by a given amount.

The aim of a cost curve is ideally to show the relationship between costs and GHG reduction (or carbon stored) over a wide range where both small GHG emission reductions and large reductions are measured. Thus, a cost curve must by its nature be an aggregate of many different technical and structural changes in the energy system. The problems inherent in describing this by one single curve are discussed in more detail in Chapter 3.

When investigating the specific cost of mitigation, i.e., cost per tonne of GHG reduced, it is important to distinguish between average, incremental, and marginal cost. The average cost of GHG reduction reflects the difference in total energy system costs when a specific mitigation scenario is compared to the baseline, divided by the difference in emissions between the two scenarios. The incremental cost can be defined as the increase in costs per unit of emission reduction when a specific mitigation scenario is compared to the previous scenario on the cost curve (rather than the baseline, as in the case for average costs). Thus incremental costs show how expensive each additional step becomes per unit of extra GHG reduced (see Figure 2.4). The marginal cost of GHG abatement is the cost of not emitting the last unit of GHG from the system, or the cost of the last unit of carbon sequestered. In an optimization model, this unit is by definition the most expensive one. The marginal cost will in this case be equal to the shadow price of the imposed emission constraint (if any), or if a carbon tax is introduced in the model, the marginal cost is equal to the tax level.

The disadvantage of only presenting average costs is that increasing costs associated with increasing emission reduction levels are leveled out (see Figure 2.4). An incremental cost curve gives a better picture of the cost consequences of each additional step of measures. Assuming that the average cost curve is increasing and convex, the incremental costs will always be equal to or higher than average costs. The marginal cost curve will then in turn always be equal to or higher than the incremental cost curve. The more model runs (mitigation scenarios) that are used to construct the incremental cost curve, the closer it will reflect the marginal cost curve. In non-optimization models, the marginal costs are not directly calculated. However, by carefully selecting mitigation scenarios when establishing the incremental cost curve, this can be used as approximation of the marginal costs.

Another issue to consider when establishing cost curves is the representation of time dependence in the costing of mitigation options. Dynamic models include linkage between time periods over the time horizon considered, and thus allow for studying the development over time of different variables, and also allow for a time-dependent description of technologies. In a dynamic model, the choice of mitigation options in one year (or time period) will depend on the choices made in previous years.[2] Static models only give a "snapshot" representation of the costs, with no time dependency included.

When interpreting cost curves it is important to be aware of what the costs presented include. Cost curves derived from bottom-up models address direct technological costs but typically ignore non-technical market factors and cost impacts of structural changes. Hence, the costs calculated in these models do not reflect GDP losses, as is the case with macro-economic models. The use of cost curves in energy-sector analysis is further discussed in Chapter 3.

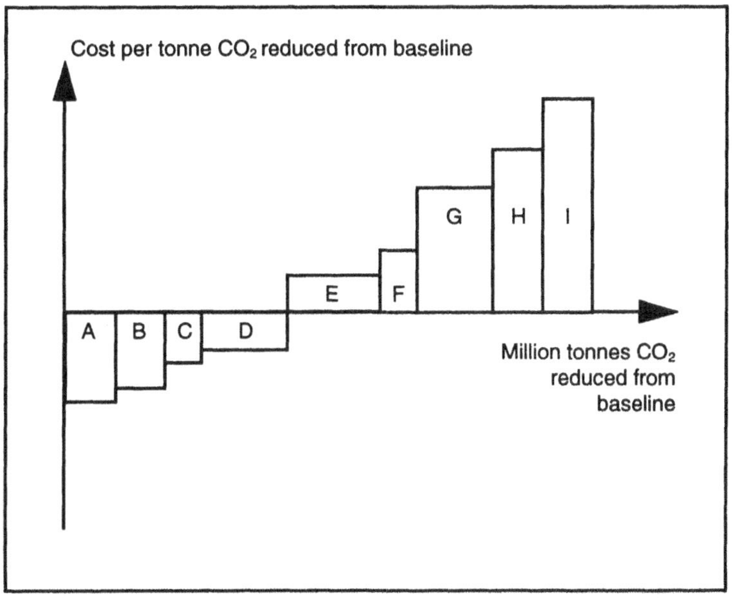

Figure 2-2. Discrete Step CO_2 - Reduction Cost Curve

[2] This is true for models with "myopic foresight". In models with "perfect foresight", the choice of mitigation options is done simultaneously for all time periods assuming that the information given to the model about both previous and future time periods is known at each time step.

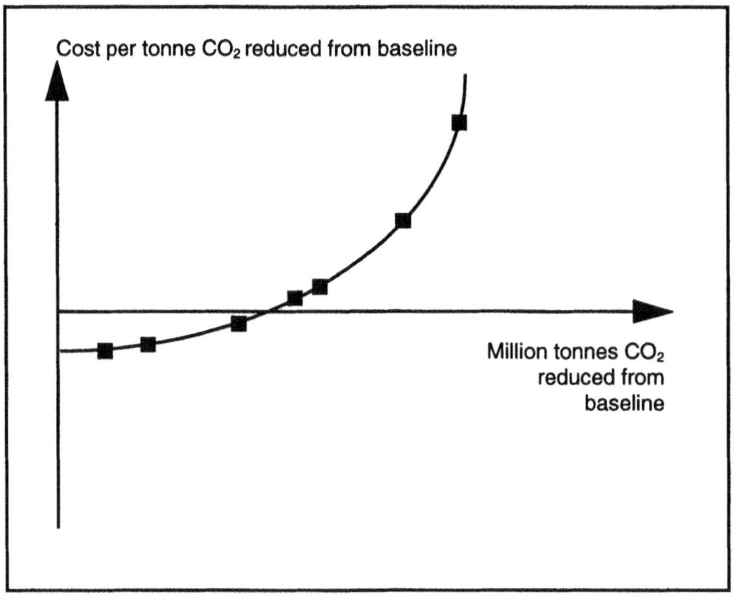

Figure 2-3. Continuous Cost Curve

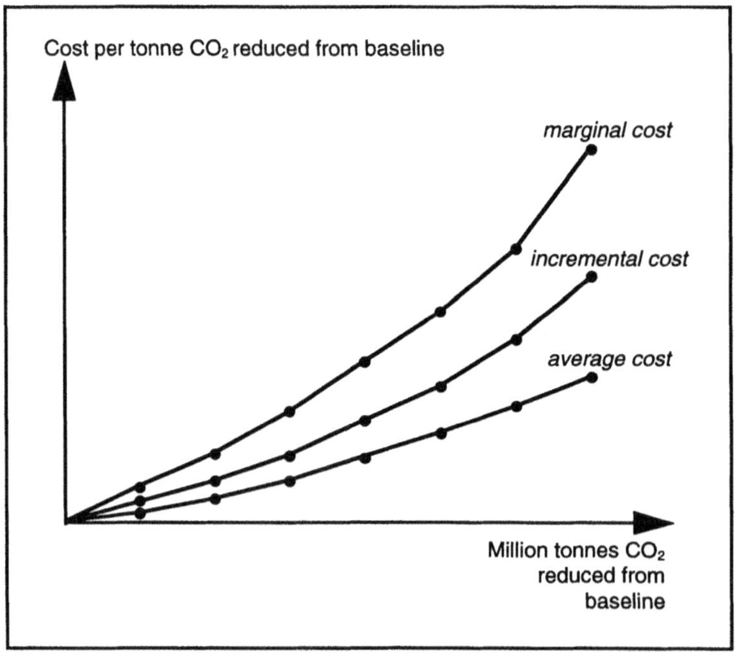

Figure 2-4. Marginal, Incremental, and Average Cost Curves

2.7.2 Accounting for Mitigation of Non-CO_2 Greenhouse Gases

Expressing the benefit of reducing emissions of different GHGs in a common unit is problematic since the effectiveness of greenhouse gases in trapping the earth's heat varies. Research on this topic has led to the development of the concept of a "global warming potential," or GWP. The GWP is intended to demonstrate the relative impacts on global warming of various gases compared with CO_2. The research conducted to date has established that the effects of various gases are too complex to permit them to be summarized in a single number. The indirect effects of some gases have proven impossible to summarize in terms of GWPs, while the direct effect depends on the time horizon considered (since gases have different lifetimes in the atmosphere). The currently available numerical estimates of GWP relative to CO_2 are given in Table 2-3 for the most important gases.

From a practical standpoint, the main issue is how to compare measures that affect methane with those that affect CO_2 and how to aggregate emissions among GHGs. One option is to present the GWP of methane mitigation options in terms of a range of estimated GWPs. For example, if the chosen range for the methane GWP was 19-110, then the CO_2-equivalent magnitude of GHG reduction for a measure that reduced methane emissions by 100,000 tonnes would be 1.9-11 million tonnes.

Table 2-3. Numerical Estimates of Global Warming Potential Relative to Carbon Dioxide

Greenhouse Gas (GHG)	Direct Effect for Time Horizons of	
	20 Years	100 Years
Carbon dioxide (CO_2)	1	1
Methane	56-110	19-43
Nitrous oxide (NO_x)	290	320
CFC-11	5,000	3,900
CFC-12	8,000	8,300
HCFC-22	4,300	1,600
HFC-134a	3,100	1,200

Source: IPCC (1992)

2.8 INTEGRATION OF ENERGY AND NON-ENERGY ASSESSMENTS

Integration of results from assessments of the energy sector and the various non-energy sectors can be a challenging task in conducting a national mitigation assessment. The degree of integration that is desired depends on the goals of a national mitigation strategy and whether policy-makers want a ranking of options across sectors.

The most basic type of integration is to simply describe the GHG impacts, costs, and other effects of particular options as identified in each of the sectoral assessments. This is the approach that has been primarily used in the OECD countries for developing and reporting National Action Plans to address climate change. General impacts on the economy are then often addressed by incorporating the different sectors in a general equilibrium framework.

If a coherent ranking of specific options across sectors is desired, however, it requires careful planning of the overall assessment and development of an analytical framework to integrate results. It also requires close communication among the various analysts before, during, and after each of the sectoral assessments. Each of the sectoral studies should use common assumptions for basic macro-parameters and also have a reasonably consistent philosophy for defining a baseline scenario.

If marginal cost curves for GHG mitigation are available from the sectoral assessments, each sector's curve can in principle be combined into a national marginal cost curve. The marginal cost for achieving a specific GHG emission reduction target can then be estimated from the national curve. One can then identify options in each sector, up to the same marginal cost, that together satisfy the overall reduction target. This would provide a theoretically "least-cost" solution for a particular national reduction target, given the limitations in the analytical methods in establishing the sectoral cost curves.

A final ranking of options across sectors may be done by assembling a range of impacts, costs, and benefits for each option (such as the criteria listed in Table 2-2). Options can be assessed in a consistent fashion using a decision analysis framework that allows for a weighting of various quantitative and qualitative criteria. Such a framework allows policy-makers to define the importance that they attach to particular criteria.

2.9 INTEGRATION WITH EMISSION INVENTORIES AND VULNERABILITY AND ADAPTATION ASSESSMENTS

The analysis of mitigation options should be closely linked to and integrated with the preparation of a GHG emission inventory and any vulnerability and adaptation assessments a country is performing. The mitigation analysis should be structured to take advantage of the information generated by these assessments, particularly the emissions inventory. It may also be beneficial to present the results of the inventory, vulnerability and adaptation, and mitigation assessments in one unified document so that conclusions can be drawn about the most important implications of climate change for a country. This will also allow decision-makers to consider the tradeoffs between implementation of adaptation measures and mitigation measures in designing a country's national strategy.

Some of the obvious points of integration between the inventory, vulnerability and adaptation, and mitigation assessments are presented below.

The results of the emission inventory and vulnerability and adaptation assessment provide useful information on GHG emissions and natural resource conditions that should be reflected in the mitigation baseline scenario. The emission inventories will provide a current estimate of GHG emissions, an accounting procedure, and a format that can be helpful in preparing emission projections for future years. The vulnerability assessment will identify likely changes in agricultural, coastal, water, forests, and other resources in the future that will affect the baseline characteristics of these resources.

The inventory results will identify the sectors and sources that have the highest emissions and contributions to global warming or the degree of removal of gases by sinks. This information on the relative current importance of sources and sinks should be considered in determining the scope and emphasis of the mitigation assessment. In addition, the emission factors (and carbon uptake factors) developed and used in the inventory should be used in the mitigation assessment in evaluating the emissions or removals associated with different mitigation options. The results of the vulnerability assessment will identify possible changes in natural resource conditions and management practices that could affect the effectiveness of mitigation options. For instance, climate-induced changes in river basin flow may affect hydroelectric potential; changes in forest growth could affect the effectiveness of reforestation programs; and changes in agricultural productivity and production practices may alter strategies for reducing agricultural emissions.

The basic assumptions about population and economic growth and natural resource conditions used in the emission inventory and the vulnerability and adaptation assessment should be consistent with the assumptions used in the mitigation analysis.

REFERENCES

Boero, G., Clarke, R. and Winters, L., 1991. The Macro-economic Consequences of Controlling Greenhouse Gasses: A Survey. Department of the Environment, UK.

Cline, W., 1992. The Economics of Global Warming. Institute of International Economics, Washington, D.C.

IPCC (Intergovernmental Panel on Climate Change). 1992. Climate Change 1992: The Supplementary Report to the IPCC Assessment. Cambridge University Press, p. 56.

Krause, F. *et al.*, 1993. Cutting Carbon Emissions Burden or Benefit: The Economics of Energy-tax and Non-Price Policies. Energy Policy in the Greenhouse, Volume II, Part 1. International Project for Sustainable Energy Paths. El Cerrito, CA.

Kydes, A., Shaw, S., and McDonald, D. 1995. Beyond the Horizon: Recent Directions in Long-Term Energy Modeling. Energy, Vol. 20, No. 2.

Lind, R. *et al.*, 1982. Discounting for Time and Risk in Energy Policy. Resources for the Future, Washington, D.C.

Lohani, B.N. and Azimi, A.M., 1992. Barriers to Energy End-Use Efficiency. Energy Policy, Vol. 20, No. 6 (June).

Sathaye, J., Norgaard, R. and Makundi, W., 1993. A Conceptual Framework for the Evaluation of Cost-Effectiveness of Projects to Reduce GHG Emissions and Sequester Carbon. Lawrence Berkeley Laboratory Report LBL-33859, Berkeley, CA.

CHAPTER 3
MITIGATION ASSESSMENT OF THE ENERGY SECTOR: AN OVERVIEW

3.1 INTRODUCTION

The energy sector is comprised of the major energy demand sectors (industry, residential and commercial, transport, and agriculture), and the energy supply sector, which consists of resource extraction, conversion, and delivery of energy products. GHG emissions occur at various points in the sector, from resource extraction to end use, and accordingly, options for mitigation exist at various points.

In most countries, the energy sector will be a major focus of GHG mitigation analysis. Globally, the energy sector is the predominant source of carbon dioxide (CO_2), the most important GHG. The combustion of fossil fuels accounts for about 60% to 90% of current net anthropogenic emissions of CO_2 emissions. The energy sector is also a source of the GHG methane (CH_4) and nitrous oxide (N_2O), and of other gases that may indirectly affect the earth's climate, such as nitrogen oxides (NO_x), carbon monoxide (CO), and non-methane hydrocarbons (NMHC). The production and transmission of coal, oil, and natural gas are estimated to account for about one-fifth of global methane emissions (IPCC, 1992).

The primary focus in Chapters 3-9 of this book is on "bottom-up", least-cost analysis. This approach involves the development of scenarios based on energy end uses and evaluation of specific technologies that can satisfy demands for energy services. One can compare technologies based on their relative cost to achieve a unit of GHG reduction and other features of interest. This approach gives equal weight to both energy supply and energy demand options. A variety of screening criteria, including indicators of cost-effectiveness as well as non-economic concerns, can be used to identify and assess promising options, which can then be combined to create one or more mitigation scenarios. Mitigation scenarios are evaluated against the backdrop of a baseline scenario, which simulates the events assumed to take place in the absence of mitigation efforts. Mitigation scenarios can be designed to meet specific emission reduction targets or to simulate the effect of specific policy interventions. The results of a "bottom-up" assessment can then be linked to a "top-down" analysis of the impacts of energy sector scenarios on the macro-economy.

3.2 STRUCTURE OF AN ENERGY SECTOR MITIGATION ASSESSMENT

Mitigation analyses require physical and economic data about the energy system, GHG emissions, socio-economic variables, and specific technology options. Using these data, a model or accounting system of the energy sector should be designed to suit local circumstances.

The manner in which a mitigation assessment is performed will reflect each country's resources, objectives, and decision-making process, and also the type of modeling approach employed. Figure 3-1 depicts the basic steps of a typical mitigation assessment and how they relate to one another. Some of the steps are interlinked, so they are not necessarily sequential, and require iterations.

An initial step is to assemble data for the base year on energy demand by sector, energy supply by type of energy source, and energy imports and exports. The disaggregated energy data should be normalized to match the national energy supply totals for the base year. One should then calibrate base year emissions with the existing GHG inventory, as available. The analyst should also assemble data for the base year on the technologies used in end-use sectors and in energy supply.

The data for the base year is used as a starting point for making projections of future parameters and developing integrated scenarios of energy demand and supply. On both the demand and supply sides, one should identify and screen potential technology options to select those that will be included in the analysis. The screening should be guided by information from an assessment of energy resources, as well as the potential for energy imports and exports.

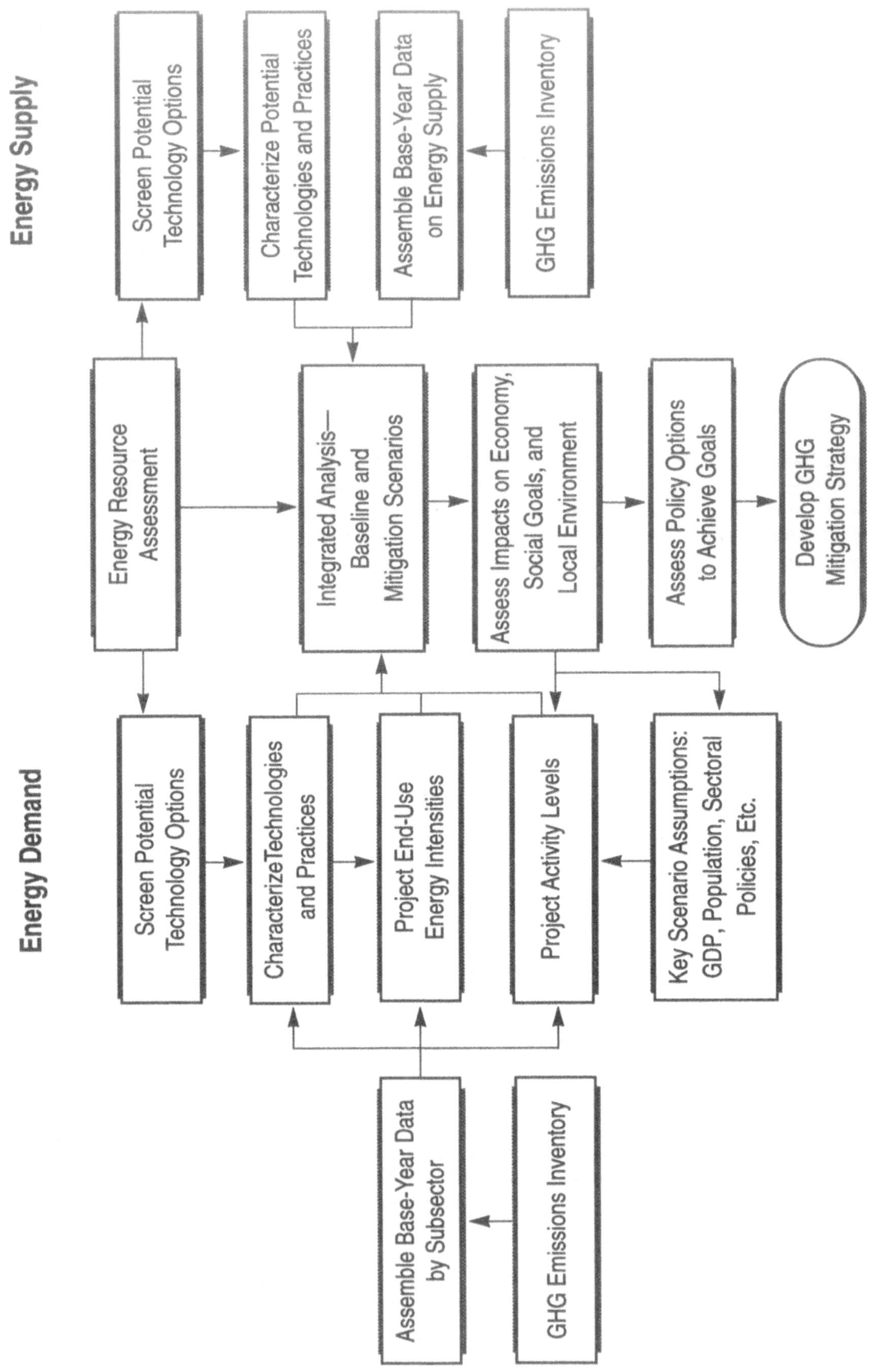

Figure 3–1. Structure of an Energy Sector Mitigation Assessement

Once the list of technologies has been made manageable by the screening process, the analyst must characterize potential technology options in end-use sectors and in energy supply with respect to costs, performance, and other features. On the demand side, this characterization will assist in projecting end-use energy demands in the various sectors. Projecting energy demand also requires one to project activity levels in each subsector for indicators such as tonnes of various industrial products, demand for travel and freight transport, and number of urban and rural households. These projections should be based on the assumptions for growth in key parameters such as GDP and population. Assumptions about sectoral policies with respect to energy pricing and other factors are also important in developing projections of energy demand.

The data from the energy demand and supply analyses is then entered into an energy sector model or accounting framework that allows for integrated analysis of the various options that can meet energy requirements. This analysis should calculate costs and impacts over the time horizon considered, review the results for reasonableness, and account for uncertainty. This step involves combining technology options to meet the objectives of each scenario. The selection of technologies may be made directly by the analyst or performed by the model (as with an optimization model).

The baseline scenario projects energy use and emissions over the time horizon selected, reflecting the development of the national economy and energy system under the assumption that no policies are introduced to reduce emissions. The baseline scenario must include sufficient detail on future energy use patterns, energy production systems, and technology choices to enable the evaluation of specific mitigation options. Alternative baseline scenarios can be developed if desired (for example, to reflect different assumptions about GDP growth).

Mitigation scenarios can be defined to meet particular emission reduction targets, to assess the potential impact of particular policies or technologies, or to meet other objectives. The comparison of mitigation and baseline scenarios should reveal the net costs and impacts of the mitigation options. The results need to be assessed with respect to reasonableness and achievability, given barriers to implementation and the policy instruments that might be used, such as taxes, standards, or incentive programs.

For both baseline and mitigation scenarios, the analyst should assess the impacts on the macro-economy, social goals (such as employment), and the national environment. One approach is to integrate bottom-up assessment with a macroeconomic model. Decision analysis methods that allow for consideration of multiple criteria may also be appropriate.

Once attractive mitigation options have been identified and their impacts characterized, the analyst should assess policy options to encourage their adoption. In practice, some consideration of policy options would likely take place as part of development of mitigation scenarios. A mitigation scenario might assume that particular policies will be implemented in order to bring about certain results. These could range from efficiency standards on end-use equipment to policies regarding power sector investments.

After scenarios have been analyzed and options have been ranked in terms of their attractiveness (in both quantitative and qualitative terms), it is desirable to conduct a more detailed evaluation of policies that can encourage adoption of selected mitigation options. Such an evaluation can play an important role in the development of a national mitigation strategy. The latter step requires close communication between analysts, policy-makers, and other interested parties.

3.3 MODELS FOR ENERGY SECTOR MITIGATION ASSESSMENT

Assessment of opportunities for GHG mitigation in the energy sector requires an accounting or modeling framework to capture the interactions between technologies and to ensure consistency in the assessment of energy, emission, and cost impacts. Accounting and modeling methods can vary greatly in terms of their sophistication, data intensiveness, and complexity. This section provides an overview of key concepts and capabilities of a number of models that are available for mitigation analysis. These models have already been used in energy/environmental studies in various developing countries.

As discussed in Chapter 2, it is common to divide energy-economy models into two types, so-called "bottom-up" and "top-down," depending upon their representation of technology, markets, and decision-making. In practice, there is a continuum of models, each combining technological and economic elements in different ways. At one extreme are "pure" energy models, which focus upon fuels and energy conversion or end-use technologies and treat the rest of the economy in an aggregated fashion. At the other extreme are "pure" economic models, which treat energy markets and technologies in an aggregated manner and focus instead upon economy-wide supply-demand relations and optimizing behavior. Between these two types are a number of models that combine elements of both extremes with various degrees of emphasis and detail.

The description of the future varies among the models. Some models can only analyze a "snapshot" year, and compare this to another year, without any representation of the transition between them. Dynamic models, on the other hand, allow for time-dependent descriptions of the different elements of the energy system. While the "snapshot" models enable great detail in the representation of the system, dynamic models allow for representation of technology capacity transfer between time periods and thus time-dependent capacity expansion, time-dependent depletion of resources, and abatement costs as they vary over time.

In dynamic modeling, information about future costs and prices of energy is available through two diametrically different foresight assumptions. With "myopic foresight", the decisions in the model are made on a year-by-year basis, reflecting the assumption that actors expect current prices to prevail indefinitely. Assuming "perfect foresight", the decisions at any year are based on the data for the entire time horizon. The model thus reflects the activities of market participants as if they use the model itself to predict prices.

The key design features of the models described in this book are summarized in Table 3-1. Most of the models listed in Table 3-1 can be used to integrate data on energy demand and supply. This information can be used by the models for determining an optimal or equilibrium mix of energy supply and demand options. The various models use cost information to different degrees and provide for different levels of integration between the energy sector and the overall economy.

In addition to the models listed in Table 3-1, we briefly describe a decision analysis methodology known as the "Analytical Hierarchy Process." It can be used to evaluate and rank options according to a combination of quantitative and qualitative criteria.

3.3.1 Energy Accounting Models

Energy accounting models such as LEAP and STAIR reflect an engineering or "input-output" conception of the relations among energy, technology, and the services they combine to produce. This view is based on the concept of energy services that are demanded by end users. Schematically, this can be represented as:

Energy inputs → *Technology* → *Energy services* →

For policy purposes, the significance of this approach is that a given type and level of energy service can be obtained through various combinations of energy inputs and technologies. In particular,

Table 3–1. Key Design Features of Selected Sector Models

Model Characteristics	STAIR	LEAP	ETO	MARKAL	ENPEP	MARKAL–MACRO	LBL–CGE
Model Type	**Energy Accounting**	**Energy Accounting**	**Engineering Optimization**	**Engineering Optimization**	**Iterative Equilibrium**	**Hybrid**	**"Top–Down"**
Number of Non-Energy Sectors	0	0	0	0	–	1	2+
Energy Supply Representation	Process Analysis	Process Analysis	Process Analysis	Process Analysis	Supply Curves	Process Analysis	Supply Curve
Energy Demand Representation	Exogenous	Exogenous	Exogenous	Exogenous	Exogenous	Utility Maximization	Demand Function
Multi-Period	No	No	No	Yes	Yes	Yes	Yes
Consumer/Producer Foresight	Not Applicable	Not Applicable	Not Applicable	Perfect or Myopic	Myopic	Perfect or Myopic	Myopic
Solution Algorithm	Accounting	Accounting	Linear Programming	Linear Programming	Iteration	Non-Linear Optimiz.	Iteration

holding the service constant while increasing the energy efficiency of the technology allows decrease in the required level of energy input. In a range of cases, when other factors are held equal, this lowers the overall cost of the energy service. In addition, the lower energy input requirement in general means lower levels of GHG emissions. With accounting models, the evaluation and comparison of policies are performed by the analyst external to the model itself.

These models are essentially elaborations upon the following accounting identity describing the energy required for satisfying a given level of a specific energy service:

$$E = AI$$

where E indicates energy, A activity, and I intensity. With multiple end uses, aggregate energy demand is simply the sum:

$$E = \sum_i A_i I_i$$

Accounting models are basically spreadsheet programs in which energy flows and related information such as carbon emissions are tracked through such identities. The interpretation of the results, and the ranking of different policies quantified in this manner, is external to the model and relies primarily on the judgment of the analyst.

Using such a model, a stylized typical analysis might proceed as follows. Using the notation above, suppose that I_i and I_i^* represent the energy intensities of current and more energy-efficient technologies, respectively. If the incremental capital and maintenance cost of shifting to the efficient technology is C for each I and the (levelized) price of energy is q, then the energy savings available from switching to the efficient technology is estimated as

$$\Delta E = \sum_i A_i (I_i - I_i^*)$$

and the associated cost saving is

$$\Delta C = \sum_i A_i [q(I_i - I_i^*) - C_i]$$

Note that these calculations assume that a number of factors are held constant, including energy service level and equipment saturations. Also, these expressions represent a "quasi-static" view; in actual practice, such calculations would be performed over time paths of costs, activities, intensities, and prices developed in scenario construction. Finally, it is easy to see that factors for carbon savings from the shift to efficient technologies could be easily included in such calculations.

In energy accounting models, macroeconomic factors enter only as inputs in deriving demand-side projections; that is, there is no explicit representation of feedback from the energy sector to the overall economy. While different models contain different levels of detail in representing the supply sector, supply-demand balancing in this type of model is accomplished by back calculation of supply from demand projections.

The energy accounting models STAIR (Sathaye *et al.* 1989), developed at the Lawrence Berkeley Laboratory, and LEAP (SEI-B, 1993), developed at the Tellus Institute, have both been widely used in developing countries to conduct energy planning, investigate carbon mitigation strategies, and study other environmental policy problems. Associated with LEAP is the Environmental Database, which contains extensive information on energy/environmental linkages.

The STAIR and LEAP models can simulate the effect of selected mitigation options on overall costs and emissions. In contrast to optimization models, they cannot easily generate a least-cost mitigation solution. They can be used to represent cost-minimizing behavior estimated by the analyst, however. They tend to require less data and expertise, and are simpler and easier to use than optimization models. See Appendices A and B for further discussion.

3.3.2 Engineering Optimization Models

In engineering optimization models, such as ETO and MARKAL, the model itself provides a numerical assessment and comparison of different policies. These models are linear programs in which the most basic criterion is total cost of providing economy-wide energy services under different scenarios; when this criterion is used, the structure of this type of model as used in mitigation analysis can be represented schematically as:

Minimize total cost of providing energy and satisfying end-use demand subject to:

- energy supplied ≥ energy demanded
- end-use demands satisfied
- available resource limits not exceeded

In addition to the overall optimization structure of these models, the key distinction between these and the accounting models is that, within the model structure itself, trade-offs are made among different means of satisfying given end-use demands for energy services.

There is variation in the intertemporal structure of these linear programming models. Some are constructed to perform a "target year" analysis: the model is first parameterized and run for the base year, then the procedure is repeated for a single designated future year (typically 2005 or 2010). Others perform a more elaborate dynamic optimization in which time paths of the parameters are incorporated and the model generates time paths of solutions.

In engineering optimization models, macroeconomic factors enter in two ways. First, they are used to construct forecasts of useful energy demands. Second, they can be introduced as constraints. For example, the overall cost-minimization can be constrained by limits on foreign exchange or capital resources. In both cases, the models do not provide for the representation of feedbacks from the energy sector to the overall economy.

Supply and demand are balanced in engineering optimization models by the presence of constraints, as indicated above. The engineering detail and level of disaggregation used in both the supply and demand side are at the discretion of the user, and in practice these vary widely among models.

This type of model allows several means of analyzing GHG emissions and the effects thereupon of various policy options. For example, as an alternative to minimizing energy costs, criteria such as minimizing carbon output subject to the constraints can be employed. In addition, an overall cap on carbon emissions can be entered as a constraint in the model, and the cost-minimization performed with this restriction. Each such approach allows the comparison of different policy intervention.

ETO was developed in India and at the Lawrence Berkeley Laboratory, while MARKAL was developed at Brookhaven National Laboratory. ETO has been applied to analyses of energy efficiency and carbon reduction scenarios in India, Brazil, and China (Mongia *et al.* 1991, La Rovere *et al.* 1994). MARKAL has been applied to energy and environmental analysis by over thirty users in twenty countries, with new countries currently undertaking its development (Manne *et al.* 1992).

3.3.3 An Iterative Equilibrium Model

The Energy and Power Evaluation Program (ENPEP) model, developed at Argonne National Laboratory, incorporates the dynamics of market processes related to energy via an explicit representation of market equilibrium, that is, the balancing of supply and demand (Buehring *et al.* 1994). ENPEP is used to model a country's total energy system and does not explicitly include an economy model integrated with the energy system model. Thus, macroeconomic factors enter the model exogenously, as in the previous model types discussed. (That is, demands for energy services are derived from macroeconomic drivers rather than being obtained endogenously.) ENPEP thus occupies an intermediate position between engineering, energy-focused models, and pure equilibrium models.

ENPEP has been used to do total energy system analysis and electric sector studies in a wide range of developing countries. It is organized in modular form for flexibility and ease-of-use. Among the modules are a library of technical data on electricity generation technologies, a detailed representation of a specified electric power sector, and module for separating electricity demand from overall energy demand. Thus, depending on the application, ENPEP's representation (particularly as regards the electric generation sector) can be quite detailed. The methodology employed to solve the model is a "process network" wherein individual energy processes are represented with standard model forms, but application-specific data, and linked together as appropriate. Prices and quantities are then adjusted iteratively until equilibrium is achieved. This iterative approach makes it much easier to include non-competitive-market factors in the system than in the optimization approach.

3.3.4 Hybrid Models

In hybrid models, such as MARKAL-MACRO and ETA-MACRO, the basic policy measure is the maximization of the present value of the utility of a representative consumer through the model planning horizon. Constraints are of two types: macroeconomic relations among capital, labor, and forms of energy, and energy system constraints. The model generates different time paths of energy use and costs and macroeconomic investment and output. The energy sub-model contains representations of end-use technologies, with different models containing different levels of detail. Schematically, this type of model can be represented as:

Maximize (discounted) utility of consumption subject to:

- macroeconomic relations among output, investment, capital, labor, and energy;
- energy system and resource constraints (as in engineering optimization models).

The constraints in this case are also dynamic: they represent time paths for the model variables.

In this type of model, energy demands are endogenous to the model rather than imposed exogenously by the analyst. In addition, this optimization structure indicates the difference we noted above in the way the different models incorporate macroeconomic data. Specifically, in accounting and engineering optimization models, these data—on GDP, population growth, capital resources, etc.—enter essentially in the underlying constructions of the baseline and policy scenarios, as discussed in previous chapters. In the hybrid model, however, such data enter in the "macroeconomic relations" (technically, the aggregate production function) as elasticities and other parameters. Within this model framework, changes in energy demand and supply can feed back to affect macroeconomic factors. It should be noted that, despite their inclusion of engineering optimization sub-components, these models typically do not contain as much detail on specific end-use technologies as many purely engineering models.

The ETA-MACRO model has been used for energy-environmental planning in a number of different countries and provides the core of the regional energy-economy modules included in the Manne-Richels Global 2100 carbon emission analysis model (Manne and Richels 1992). This model combines an aggregate process analysis model of electric and non-electric energy production with a aggregate economy-wide production function with labor, capital, electric, and non-electric energy inputs. The strength of this model is its ability to represent a number of fundamental energy-economy relationships with a minimal set of input data. In addition to cost data on electric and non-electric generation, the model only requires a few macroeconomic inputs (e.g., labor force participation and productivity growth rates), resource inputs (e.g., oil and gas resources), and macroeconomic behavioral parameters (e.g., the rate of substitution between energy and other inputs into the economy).

MARKAL-MACRO, developed at Stanford University and the Brookhaven National Laboratory, also provides an integrated representation of macroeconomic relations and energy technology processes (Manne *et al.* 1992). This model, however, contains much greater detail in the end-use or process analysis component. It can address conservation and energy-efficiency changes in end-use devices directly. The model-wide objective function combines with the cost-minimization in the linear programming sub-model, which ensures that energy demands are satisfied at least cost (both on the supply and the demand sides). This model represents the most complete effort to-date to combine the "top-down" and "bottom-up" approaches. MARKAL-MACRO has been applied in the United States and is currently being used in ten countries participating in the International Energy Agency's Energy Technology Systems Assessment Programme.

3.3.5 "Top-down" Models

Top-down models focus on economic equilibria, with less emphasis upon details of energy technology and end-use analysis. They are built upon the assumptions of competitive equilibrium and optimizing behavior on the part of consumers and firms. They differ from hybrid models in being based upon general equilibrium rather than neoclassical growth theory, and in eschewing detailed engineering sub-components for energy system representation. While theoretically capable of including deviations from these assumptions in the form of representations of "imperfect competition," such developments have not been pursued in the context of energy analysis.

The MIMEC model has been used to study development, energy, and environmental policy in Egypt and India (Eckhaus and Lahiri 1994). A distinctive feature of this model is the use of a process analysis form of input-output analysis to summarize the production possibilities of the economy. In the two applications, ten sector (Egypt) and eighteen sector (India) input-output tables were constructed using historical data; various substitution elasticity estimates were used to develop alternative input vectors for producing the output of the sector in question; and input vectors for alternative technologies (those that have been introduced, but are not yet in widespread use) were included directly. The selection of technology for the production of the output of each sector depends on the relative costs of the inputs to it. The MIMEC model is capable of addressing information on costs and energy savings by technology type for selected industries. Thus, in-depth analysis of selected industries can be linked through this model to overall economic activity. The analysis of other sectors cannot be explicitly represented in this model. The model chooses a mix of products that maximizes the discounted utility of consumption of a representative consumer with a linear expenditure system underlying the model of consumer behavior.

The LBL-CGE model is a multi-sector computable general equilibrium model designed for analyzing energy and environment policies. It focuses on macroeconomic effects of investment in the energy sector and also on effects of energy price increases on energy consumption in different sectors of the economy. The above are contrasted with effects of investment in improving energy efficiency. The energy/economy interaction is modeled by dividing the economy into materials and energy sectors. The materials sector is subdivided into agriculture, basic industry, other industry, construction, transport, and the services sector. The energy sector is divided into crude extraction, natural gas, electricity and petroleum products. The model allows for interfuel substitution both in production activities as well as in household consumption. Overall energy consumption is assumed to be sensitive with respect to prices and investment in energy efficiency. The effects of policies are analyzed on different points of time in a comparative static framework. Unlike MIMEC, continuous functions rather than process analysis are used to represent production possibilities. It is proposed to experiment with alternative functional forms and nesting schemes for modeling production technology.

3.3.6 Decision Analysis Methodology

The Analytical Hierarchy Process, or AHP, is a decision analysis tool that allows the explicit application of both quantitative and qualitative policy criteria. It includes a technique for weighing different criteria in combinations according to the judgment of the analyst. Possible criteria include cost (life-cycle and capital), resource availability, social acceptance, state-of-development, environmental impact, and infrastructure requirements. The AHP has been applied by the U.S. National Renewable Energy Laboratory to the study of renewable energy technologies in Mexico.

3.4 KEY STEPS IN THE MODELING PROCESS

The modeling process typically consists of a number of interrelated steps. Seven key steps in the process are briefly discussed here: (1) determining model objectives, (2) model formulation, (3) data collection, (4) data analysis, (5) model calibration, (6) model implementation, and (7) sensitivity analysis.

The first step in the model building process is typically the selection of one or more objectives for the analysis. For example, if carbon emissions are to be addressed in a model, are costs and benefits of carbon emissions to be balanced? Are emissions to be reduced a specified amount at minimum cost? Are the impacts of various policies on carbon emissions to be traced? If benefits, costs, or impacts are to be considered, how are they to be measured?

The choice of objective(s) places some restrictions on the model formulation, but a number of alternatives are still generally feasible. Should the model be formulated to optimize some objective, to compute an energy, or economy-wide equilibrium, or simply to simulate the evolution of the market as various external forces and policies evolve? Crucial decisions made here include the amount of geographic, and sectoral aggregation, the time horizon, the disaggregation of time into time periods, the amount of foresight to attribute to various economic actors, and the decision rules used by those actors.

These considerations influence the decision among energy accounting, engineering optimization, and economic optimization models. In the accounting models, the comparison among policies, including cost-benefit calculations, is carried out externally by the analyst. In the engineering optimization models, minimum costs associated with different energy sector configurations are computed by the model, but these are not explicitly integrated with the larger economy. The hybrid models provide integrated optimization within the model itself, thereby enabling the analyst to rank policies directly.

Steps three and four address data collection and analysis. Model formulation is often constrained by the amount and quality of the data available. Two types of data may be employed: (1) historical economic data regarding economic activity, amounts of labor, capital, materials, and energy consumed, and (2) engineering data concerning the costs of existing and likely future technologies. Once data has been collected it may be analyzed in a number of ways. It may simply be checked for quality, or statistical methods may be used to estimate relationships between key economic parameters.

Data considerations also differ among different types of models. In the more engineering-focussed models, input-output data on energy use and efficiency can be the primary need, with only aggregate macroeconomic data required to derive forecasts. In models that employ more detailed economic elements, estimates of parameters such as elasticities of substitution are also required. The latter can dictate more sophisticated statistical methods for model implementation.

When the necessary data have been introduced into the model, it is then usually necessary to calibrate it to base year conditions. Of course, if credible independent estimates of enough of these parameters are available, care must be taken to calibrate by varying the least reliable parameter values.

Once the model has been calibrated, it is usually embedded in a user-friendly computer environment to allow inputs to be changed and outputs observed easily, and the model to be solved efficiently. Once the model has been implemented, extensive sensitivity analyses are usually run to determine to which inputs the model outputs of interest are most sensitive. This yields information of use to policy-developers and gives the user more or less confidence in the model, depending on how the results square with intuition. Sensitivity analyses, thus, form an important part of the model assessment process, which also includes third-party review of the data and formulation of the model and comparisons of the results it produces with those produced by other models with the same objectives.

3.5 KEY CHALLENGES IN THE MODELING PROCESS

A number of key challenges must be met in order to proceed through the model-building process successfully. Each modeling project confronts its own unique set of challenges, but some appear to be common to almost any effort. These include (1) incorporating efficiency versus equity; (2) aggregation over time, regions, sectors, and consumers; (3) representing decision rules used by consumers; (4) incorporating technical change; (5) capturing facility retirement dynamics; (6) avoiding extreme solutions; and (7) accounting for carbon flows.

None of the models we have discussed provides explicitly for making trade-offs between efficiency and equity. Different models, however, have different implications for the analyst's consideration of this important issue. Non-optimization models do not themselves choose among different policies in an explicit way, but can allow for the ranking of policies according to criteria specified by the analyst, including considerations of equity. Engineering optimization models, since they focus on least-cost solutions to the provision of energy services, leave to the analyst the judgment of how to trade off the importance of energy with that of other economic and social priorities. The models which optimize the utility of a representative consumer, in a sense, constrain consideration of the issue the most. Embedded in this modeling structure is a view of the economic system that equates social optima with competitive economic equilibria; the appropriateness of this perspective in the application at hand must be weighed in selecting a model.

Perhaps the most fundamental formulation issues faced by the model-builder involve the level of aggregation at which costs and benefits are calculated. Economic efficiency is generally insured if discounted benefits less costs are maximized in the aggregate and any desired income re-distribution is handled subsequently. The models described here do not explicitly examine equity considerations; these would have to be examined outside the model. Decision-makers, however, are fundamentally interested in how the costs and benefits fall on various income, industry, and regional groups. Coupled with the relative emphasis of the analysis on equity versus efficiency is the desired level of disaggregation of the model by region, time periods, industry, and income group. Obviously, the level of disaggregation must be sufficient to allow reporting of results at the desired level. If possible, it is important to capture the heterogeneity in decision-making on the part of the different groups. Decision rules are critical elements of the models and range from minimizing discounted future costs (or maximizing benefits) over a 40- or 50-year time horizon to picking investments that come close to minimizing costs based on conditions for a single year only (which is more typical of the LBL-CGE model).

Accounting models contain no explicit representation of consumer decision-making. In practice, however, their use often reflects the view that certain "market barriers" constrain consumers from making optimal decisions with respect to energy efficiency. At the other extreme, the use of the "representative consumer" in the optimization models rests upon very strong assumptions regarding consumer behavior. Key among these is the assumption of perfect foresight. It is also important to note that this method of representing the aggregate outcome of individual decision-making serves primarily to ensure mathematical and empirical tractability, and has only weak theoretical justification as a model of market processes.

Another set of key assumptions about inputs are those made about the costs and efficiencies of current and future technologies, both for energy supply and energy use. Most analysts use a combination of statistical analysis of historical data on the demand for individual fuels and a process analysis of individual technologies in use or under development to represent trends in energy technologies. At some point these two approaches tend to look similar, as the end-use process analysis usually runs out of new technology concepts after several decades, and it is then assumed that the efficiency of the most efficient technologies will continue to improve as time goes on. Particularly important, but difficult, here is projecting technological progress. Attempts to empirically estimate the dependence of future trends in productivity on factor prices, including energy, are rare.

Most current modeling approaches focus on investments in new energy-producing and -consuming equipment, which is typically assumed to have a fixed useful lifetime. In scenarios where conditions change significantly (either through external factors or explicit policy initiatives), it may be economical to retire facilities earlier or later than dictated purely by physical depreciation rates. This endogenous calculation of facility retirement dates can be handled analytically in most models, but represents a major increase in data and computational requirements.

Another typical problem, particularly with models that assume optimizing behavior on the part of individual economic agents, such as MARKAL-MACRO and MIMEC, is the danger of "knife edge" results, where a few cents difference in the cost of two competing technologies can lead to picking the cheaper one. This is generally handled by disaggregating consumers into different groups who see somewhat different prices for the same technology (e.g., coal is cheaper in the coal fields than a thousand miles away), modeling the decision process as somewhat less than perfect, and/or building appropriate time lags into the modeling structure.

Finally, estimating carbon flows for a given energy system configuration can be complicated. It is more accurate to measure emissions as close to the point of combustion as possible so types of coal and oil products can be distinguished and feedstocks (which do not necessarily produce carbon emissions) can be netted out. However, a point of use model requires far more data and computation than the models described here which aggregate several fuel types and use average carbon emissions factors for each fossil fuel.

3.6 SELECTION AND USABILITY OF MODELS FOR MITIGATION ASSESSMENT

The selection of an appropriate model depends on the purpose for which the model will be used and the data and human resources available to make proper use of it. It is important to list the questions that the model will be used to address, in order of priority, prior to selecting a model.

The models described in this chapter differ in their application for mitigation studies among countries. Several usability criteria appear to be critical, but how critical each criterion is depends on the analytic environment in the country under consideration, e.g., Ph.D. economists may be plentiful in some countries but scarce in others. Table 3-2 shows how the models considered in this chapter can be scored on several usability criteria, which are further discussed below.

Data requirements. The selection of a model for mitigation assessment needs to be guided by the data that are already available and those that might be collectable within the time period of the assessment. A mitigation assessment requires that data be collected for various energy forms and on the technical characteristics of mitigation measures. Switching to less-carbon-intensive fuels and more efficient supply and use of energy are two basic types of mitigation opportunities. Switching includes the use of renewables. The availability of data for both renewables and energy-efficiency opportunities is inadequate in most developing and transitional countries, as official data collection is primarily focused on conventional fuels.

An energy accounting model (STAIR or LEAP) has the simplest framework which can be used for mitigation assessment with the minimum of non-energy data. But even in this case, if a mitigation assessment requires the evaluation of costs, then data on cost of individual options is needed. Most countries have energy balances, but a mitigation assessment will need data well beyond these, and an analyst needs to bear this fact in mind before attempting a detailed mitigation assessment.

The other bottom-up models (ETO, MARKAL and ENPEP) require increasingly more sophisticated data sets. Since the optimization models rank options on the basis of their costs of providing energy services, reliable estimates of costs become crucial to the least-cost selection of technology options. The ENPEP model complicates this further by requiring both supply and demand curves for each type of fuel. Gathering data for these curves can be difficult when future resources are not known with much certainty.

The hardware and software requirements for each model can be crucial in the early stages of model development, but these may be the easiest constraints to surmount as computer costs fall and the requisite computer infrastructure (particularly as regards technical support services) develops in each country.

The ease of modifying the source code of the models varies considerably. LEAP and ENPEP use a proprietary software code and a user cannot access the model code and make changes to it. STAIR and ETO, on the other hand, use an open structure whereby the user can make changes to the code. STAIR uses either Lotus 123 or Symphony and ETO uses the GAMS software. MARKAL and MARKAL-MACRO are also programmed using GAMS, and the user can in principal access the source code and make changes to it.

The advantage of being able to change the code is that the user can change the type of output desired from the model. For instance, if the user wishes to focus on capital investment and foreign exchange in addition to estimating the costs of providing energy services, he/she can make the changes to the ETO code to serve this purpose. Changing the source code requires a detailed understanding of the structure of the software and should only be done by a skilled modeler.

Table 3–2. Ease-of-Use Factors for Selected Models

CRITERIA	STAIR	LEAP	ETO	MARKAL	ENPEP	MARKAL–MACRO	LBL–CGE
Data Requirements	Light–Moderate	Moderate	Moderate	Moderate	Moderate–Heavy	Moderate–Heavy	Moderate
Hardware Requirements	PC286	PC386	PC486	PC486	PC486	PC486	PC386
Software Requirements	Spreadsheet	Spreadsheet	GAMS–MINOS	GAMS–MINOS	Internal	GAMS–MINOS	GAMS–MINOS
Ease of Modifying Source Code	Easy	Harder	Easy	Moderate	Harder	Moderate	Easy
Ease of Modifying Data	Easy	Easy	Harder	Easy	Easy	Easy	Harder
Ease of Including Non–Market Factors	Not Applicable	Not Applicable	Harder	Harder	Easy	Harder	Easy

A disadvantage of changing the software is that the software will not be identical across all users, which complicates comparison of results. Modifications of the source code without guidance from the software provider/developer will also make support and maintenance difficult. Some of the models also have a high degree of "built-in" flexibility; for example, in MARKAL an option of providing user-defined equations is included. The modular design of ENPEP also allows the user a high degree of flexibility.

Ease of modification of input data and processing of results. Both ENPEP and MARKAL/MARKAL-MACRO are provided with menu-driven user interfaces that simplify modification of input data and structuring of the model. For both systems, demonstration cases with default data are provided. These interfaces also organize file handling and execution of runs and include result processing menus for both tabular and graphical display. A similar menu-driven data entry system is provided with LEAP, which also includes a flexible graphics reporting system.

ETO and LBL-CGE do not have any specialized user interface. In both cases, modification of data is done directly in the GAMS input file. The results are checked and processed through investigation of the output file produced by GAMS. For presentation and analysis of the results, the GAMS output file can be modified and imported to spreadsheet software like Excel.

Incorporation of non-market factors. For some countries, a significant issue can be incorporating factors other than those that lead to competitive market equilibrium solutions. Typically, optimization models can compute a competitive equilibrium market solution quite efficiently, but they must be modified, sometimes substantially, to include factors that depart from the competitive equilibrium paradigm. Models which search iteratively for an equilibrium solution are generally less efficient in finding competitive equilibria, but easier to modify to take into account non-competitive market factors. One prime example of an important non-competitive market factor is average cost pricing for electricity. The optimization models must be modified to convert the marginal cost prices they produce automatically into average cost prices, while the iterative equilibrium models can accommodate any desired pricing rule directly. The accounting models, since they are not based upon the equilibrium concept, do not run into these problems directly.

3.6.1 The Types of Questions That Each Model Can Best Address

Following are some key questions that an analyst would seek to answer in a mitigation assessment:

1. What is the economic cost of providing energy services in a baseline or mitigation scenario? What is the incremental cost between scenarios?

1a. What are the capital and foreign exchange implications of pursuing alternative scenarios?

2. What is the economic cost of pursuing particular mitigation options, such as high-efficiency lighting or a renewable technology, and what are its carbon implications?

3. What are the costs of reducing carbon emissions to a predetermined target level? (Target may be annual or cumulative.)

4. What is the shape of the marginal cost curve for reducing carbon emissions?

4a. How do alternative technologies rank in terms of their carbon abatement potential?

Any one of the three types of models discussed above can address questions 1 and 2.[1] However, only the ETO model is currently set up to address the foreign exchange component of question 1a, which is important for developing country planners.

Question 3 is easiest to address using an optimization model. The other models may require several runs to come up with an answer. Of the two optimization models, MARKAL is a dynamic model, while ETO focuses on a target year. MARKAL can evaluate the impacts of a cumulative carbon constraint, as well as annual constraints changing over time. Both models can evaluate target-year constraints.

Question 4 requires that energy supply and demand be evaluated in an integrated manner. A demand-side mitigation measure may change the energy supply configuration measure, which will affect the GHG emissions of the energy system. The optimization and iterative models are capable of capturing the integrated effect and deriving the changes in carbon emissions. The accounting models may not capture the changes in the energy supply mix and the consequent GHG emissions, and thus may show higher emissions reduction for a mitigation measure. Only the optimization models can calculate marginal costs directly. In the other models, an approximation of the marginal cost curve can be constructed by performing a large number of carefully selected runs (see discussion of cost curves in Chapter 2).

3.7 DESIGNING AND SETTING UP A BOTTOM-UP ANALYSIS

The first step in conducting a mitigation analysis is to begin the collection of relevant data. The analyst must also define several key parameters that will guide data collection and analysis, such as the base year and the time horizon of the study. Data are then assembled in a modeling or accounting framework, which should be designed to reflect local conditions and priorities. In order to ensure consistency, the detailed data should be calibrated to match official aggregate energy supply totals and any existing GHG inventory.

The exact data requirements for a given study will depend on the nature of the energy demand and supply model that the analyst develops. Development of the model structure and relationships, in turn, depends upon the availability of disaggregated data, local circumstances and priorities, the relative importance of sectors and end uses in terms of current and projected GHG emissions, and the specific mitigation options that will be considered. As a consequence, data collection, the specification of model structure, and the screening of mitigation options are interdependent and thus iterative tasks.

Two key considerations are the choice of discount rates and the definition of which costs and benefits to include in the analysis. Direct economic costs include the costs for equipment, operations and maintenance, and energy resources. The analyst might want to estimate the administrative and program costs associated with measures to encourage adoption of mitigation technologies and practices.[2] The analyst must also decide whether less tangible social and environmental costs will be included.

---

[1] Each of the models described above currently ranks or selects technological options, or can do so, on the basis of their net present value or annualized energy cost. The models thus assume that energy costs determine the choice among demand- and supply-side technologies and processes. In the optimization models ETO and MARKAL, technologies can also be selected on the basis of total cost of industrial production, which assumes that industries make their technology choices on the basis of total production cost rather than energy cost alone.

[2] To specify these costs more precisely would generally require that the exact mechanism or program for efficiency improvement be identified. For the purpose of an initial mitigation analysis, simple assumptions will generally suffice and can be modified once specific programs are evaluated.

3.7.1 Data Collection

Regardless of the approach taken and analysis tool used, the collection of reliable data is a major and relatively time-consuming aspect of mitigation analysis. In order to keep data constraints from becoming a serious obstacle to the analysis, two points are essential. First, modeling tools should be sufficiently flexible to adapt to local data constraints. Second, the data collection process should be as efficient as possible. Efficiency can be maximized by focusing the detailed analysis on sectors and end uses where the potential for GHG mitigation is most significant and avoiding detailed data collection and analysis in other sectors.

Data collection generally begins with the aggregate annual energy use and production figures typically found on a national energy balance sheet. The remaining data requirements will depend largely on (a) the disaggregated structure of the analysis; (b) the specific mitigation options considered; and (c) local conditions and priorities.

The typical types of data needed for a bottom-up approach to mitigation analysis are shown in Table 3-3. They tend to fall within five general categories: macroeconomic and socioeconomic data; energy demand data; energy supply data; technology data; and emission factor data. The full listing of potential data requirements may appear rather daunting. In practice, however, much of the data needed may already be available in the form of national statistics, existing analytical tools, and data developed for previous energy sector studies. The development and agreement on baseline projections of key variables, the characterization of mitigation options relevant to local conditions, and, if not already available, the compilation of disaggregated energy demand data are typically the most challenging data collection tasks facing the analyst.

In general, emphasis should be placed on locally derived data. The primary data sources for most mitigation assessments that are carried out over a period of around one year will be existing energy balances, industry-specific studies, household energy surveys, electric utility company data on customer load profiles, oil company data on fuel supply, historical fuel price series maintained by government departments, vehicle statistics kept by the transportation department, etc. The main thrust of the data collection effort is not so much on collecting new primary data but on collating secondary data and establishing a consistent data set suitable for analysis using the model of choice.

Where unavailable, local data can be supplemented with judiciously selected data from other countries. For example, current and projected cost and performance data for some mitigation technologies (e.g., high-efficiency motors or combined-cycle gas units) may be unavailable locally, particularly if the technologies are not presently in wide use. For this purpose, technology data from other countries can provide indicative figures and a reasonable starting point. For data on energy use patterns, such as the fraction of electricity used for motor drive in the textile industry, the use of external data can be somewhat more problematic. In general, it may be possible to use estimates and general rules of thumb suggested by other country studies, particularly data from other countries with similar characteristics.

Table 3-3. Data Sources for a Bottom-up Mitigation Analysis

Data Categories	Types of Data	Common Data Sources
Macroeconomic Variables		
Aggregate driving variables	GDP/value added, population, household size	National statistics and plans; macroeconomic studies
More detailed driving variables	Physical production for energy intensive materials; transportation requirements; agricultural production and irrigated area; changes in income distribution, etc.	Macroeconomic studies; transport sector studies; household surveys, etc.
Energy Demand		
Sector and subsector totals	Fuel use by sector/subsector	National energy statistics, national energy balance, energy sector yearbooks (oil, electricity, coal, etc.)
End-use and technology characteristics by sector/subsector	Energy consumption breakdown by end-use and device: e.g. energy use characteristics of new vs. existing building stock; vehicle stock; breakdown by type, vintage, and efficiencies; or simpler breakdowns	Local energy studies; surveys and audits; studies in similar countries; general rules of thumb from end-use literature; see e.g. Geller (1991); Reddy (1991); Schipper and Meyers (1992)
Response to price and income changes (optional)	Price and income elasticities	Local econometric analyses; energy economics literature
Energy Supply		
Characteristics of energy supply, transport, and conversion facilities	Capital and O&M costs, performance (efficiencies, unit intensities, capacity factors, etc.)	Local data, project engineering estimates, Technical Assessment Guide (US EPRI, 1993); IPCC Technology Characterization Inventory (US DOE, 1993)
Energy prices		Local utility or government projections; for globally traded energy products, see World Energy Council (1992); UNEP (1994a); US DOE/IEA (1994)
Energy supply plans	New capacity on-line dates, costs, characteristics	National energy plans; electric utility plans or projections; other energy sector industries (refineries, coal companies, etc.)
Energy resources	Estimated, proven recoverable reserves of fossil fuels; estimated costs and potential for renewable resources	Local energy studies; World Energy Council (1992); Johannson et al. (1993)
Technology Options		
Technology costs and performance	Capital and O&M costs, performance (efficiencies, unit intensities, capacity factors, etc.)	Local energy studies and project engineering estimates; technology suppliers; other mitigation studies (UNEP, 1994); IPCC Technology Characterization Inventory (US DOE, 1993)
Penetration rates	Percent of new or existing stock replaced per year; overall limits to achievable potential	
Emission Factors	Kg GHG emitted per unit of energy consumed, produced, or transported	National inventory assessments; IPCC Inventory Guidelines (IPCC, 1994b); IPCC Technology Characterization Inventory (US DOE, 1993)

3.7.2 Selecting and Characterizing Technology Options to Include in the Analysis

The criteria for judging whether a specific option should be included in a mitigation analysis must encompass social, political, and cultural factors in addition to standard economic concerns. GHG mitigation must be integrated with other key national objectives, such as improving the balance of payments or promoting rural development. Some of the potential criteria that could be used for evaluating mitigation options are listed in Chapter 2. The final specification of screening criteria—and weighing the relative importance of those criteria—will depend on local conditions and priorities.

Using a broad set of screening criteria to weed out unpromising, undesirable, or infeasible options requires considerable judgment. Many criteria will likely be qualitative and difficult to measure in an objective fashion. An obvious reason for screening out an option might be the infeasibility of its wide-scale application. Location of options—e.g., fuel resource options or power plants—in environmentally or otherwise sensitive areas may rule them out. There might also be overriding concerns about political acceptability. In addition, there may be options, such as reducing traffic congestion, which may be difficult to analyze since quantifying the impact on GHG emissions may be difficult to do.

Mitigation technology options can be identified from a variety of sources: country case studies, literature review, and international data bases such as those prepared by the IPCC and the International Institute for Applied Systems Analysis (IIASA) (see Box 3-1). If derived from studies in other countries, the cost, performance, and emission characteristics of specific options should be reviewed for their relevance in the local situation.

Box 3-1. Data Sources on Technologies for Reducing GHG Emissions

Two sources of information that may be useful in a mitigation assessment are the *IIASA CO_2 Technology Data Bank (CO2DB)*, developed by the International Institute for Applied Systems Analysis in Austria, and the *Inventory of Technologies, Methods, and Practices for Reducing Emissions of GHG*, prepared for the IPCC by the U.S. Department of Energy and the Environmental Protection Agency. Both of these sources contain a consistent set of information on many energy supply and end-use technologies. They each describe technical, economic, and environmental characteristics of the various technologies. The IPCC *Inventory* also describes implementation requirements (labor and infrastructure).

The *CO2DB* presently contains data on about 350 technologies, including conventional as well as more advanced technologies. The IPCC *Inventory* currently contains 43 energy supply and 44 end-use technologies, and is focused more on advanced technologies. The source country for most of the information in the IPCC *Inventory* is the U.S., while the *CO2DB* is more oriented toward European sources. To enhance the usability of the technology data bank, IIASA has developed a fully interactive software package that simplifies retrieval of information.

The key data required for each technology are life-cycle costs (including capital, fuel, and operation and maintenance), energy consumption or production features, lifetime, and environmental characteristics. For end-use technologies, costs can be entered in terms of costs of conserved energy (CCE) for efficiency improvements or the costs of specific equipment. For energy supply technologies, cost data can be entered in different formats depending on how the model is structured. These options include specifying combined costs per unit of energy produced (e.g., cost per kWh) or specific capital, fixed and variable operating and maintenance (O&M), and financing costs by plant type. It may not always be possible to enter all of the relevant data in a particular model, but it is important to have a convenient means of storing such information so it may be used in a multi-criteria assessment.

One approach for characterizing mitigation options is to employ individual experts with detailed knowledge of specific sectors. For example, a mitigation scenario analysis conducted for the U.S. involved teams of sector specialists who conducted separate analyses of technical and policy options for the buildings, transport, industrial, and energy supply sectors, using a common screening approach (UCS *et al.*, 1991). These sectoral analyses generated the cost and penetration rate estimates for selected options that were then entered into the LEAP model as data for the integrated scenario analysis.

3.7.3 Design of Model Structures

The development of demand and supply modeling (or accounting) structures is generally performed in parallel with the data collection process. The principal objectives in developing a model structure are: (a) to represent the national energy system and the major factors that influence its development, and (b) to include a sufficient level of detail to permit the analysis of selected mitigation options. For instance, where irrigated agriculture accounts for a significant share of electricity consumption and related GHG emissions, the model structure might include greater end-use detail in the agricultural sector to enable the evaluation of options such as improved pumpsets or more effective water delivery. In other countries, such model structure and data requirements might be irrelevant. In general, modeling (or accounting) structures should be tailored to local circumstances.

High levels of disaggregation and detailed data structure are characteristic of end-use approaches. While disaggregated data can help in evaluating specific technological options, excessive data disaggregation can be an analytical burden. Where local data are scarce or unreliable and are augmented by the use of secondary data or assumptions, the better resolution otherwise provided by disaggregation is lost in the fuzziness caused by the data. An efficient solution can be to use higher resolution disaggregated approaches in key sectors or subsectors where the potential for cost-effective GHG mitigation appears high. Less detailed methods can be used to provide a broader overview in other sectors.

3.7.3.1 Energy demand structure

The way in which energy demands are represented varies somewhat among different models. An example of a demand analysis structure (from the LEAP Demand program) is shown in Figure 3-2. To develop such a structure, the analyst must:

- **Design a "branch structure."** The analyst specifies the demand sectors (e.g., residential, industry, and transport) to be modeled, then breaks down each sector, to the degree desired and appropriate, into subsectors (e.g., iron and steel production, chemicals, etc.), end-uses (e.g., motive power, process heat) and devices (e.g., electric motors, furnaces, boilers, etc.) as necessary. Each of the four levels may have as many, or as few, branches as are required. Greater disaggregation should be used to model those sectors, subsectors, end uses, and devices where the potential for GHG reduction is greatest. The design of a demand structure often requires frequent references to the available data as the design progresses.

- **Specify the appropriate variables for each branch.** Here the task is to pick the relevant variables that "drive" the demand for energy. Consider the simple representation of cooking in urban households shown in Figure 3.2. At the sector level, either population or the total number of households could be specified. Since cooking generally occurs at the household level, the number of households is usually a better driving variable than population for projecting energy use for cooking. Similar choices need to be made for other branches, e.g., whether energy use in the iron and steel industry should be a function of physical production (e.g., tonnes of an indicator commodity, such as steel, produced) or economic output (e.g., value added for the industrial subsector). These choices will depend on local conditions, data availability, and robustness of the relationships.

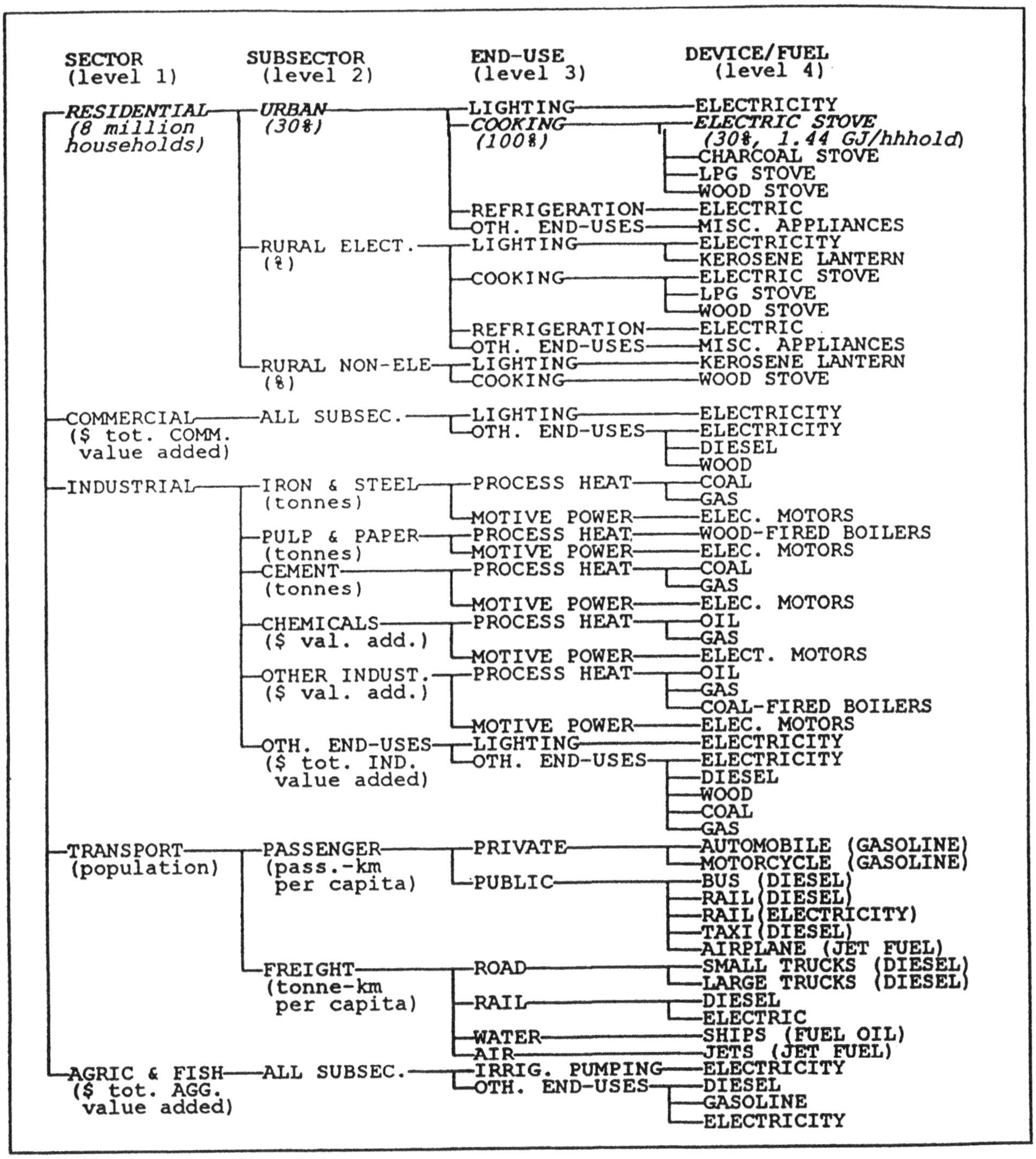

Figure 3-2. Schematic of Demand Structure for Country X (with key driving variables added in parentheses)

The urban household cooking data in Figure 3.2 illustrates the activity level and energy intensity specification for one demand "branch". The numerical data shown are for the base year, 1990. At the first level, the driving activity is households, of which there are assumed to be 8 million in 1990. At level two, the fraction of these households in the "Urban" subsector is entered (30% in 1990). At the third and fourth levels, the saturation of the cooking end use (100%, since most every household cooks) and fuel shares and energy intensities are entered, respectively. For example, 30% of urban households in 1990 used electric stoves at an annual final energy intensity of 1.44 GJ per household.

Some data values can be taken directly from existing statistical compilations, while the analyst may have to derive or estimate other values based on existing statistics, survey data, or even representative data from nearby countries. In some cases, only more aggregate statistics will be available. For example, the analyst may know the average electricity use per urban household, but only have suggestive data about the saturation and electricity use levels for specific end uses (such as lighting, cooking, or refrigeration). In such cases, the analyst will have to use judgment and refer to other studies.

3.7.3.2 Energy supply structure

The design of an energy supply structure varies considerably among different models. Optimization models, and especially a model such as ENPEP, have more complex representations of the energy supply system, allowing for a detailed description of options for energy extraction, conversion, and distribution that the model can choose among. Some general steps in designing an energy supply structure are:

- **Define the important energy extraction, conversion, and distribution activities in the country.** Energy sector models generally contain a default list of energy transformation "modules" that can be changed to suit local conditions and the mitigation options being considered.

- **Specify process data**. Within each transformation module, the analyst specifies more detailed "process-level" data describing individual facilities or types of facilities such as electricity generation plants. The degree of detail in which data are specified should reflect the importance of each energy supply sub-sector in the mitigation analysis. In many countries, the electric sector is the most important GHG emissions source so the analysis may call for more detailed data on the cost and performance characteristics of individual generation technologies. For each technology, the analyst specifies input fuels, capacity, efficiency, capacity factor, and the capital and operating and maintenance costs, and, if relevant, the co-production of other energy outputs, such as cogenerated steam.

- **Define the operating rules for each transformation sector**. The analyst indicates how facilities will operate to meet demand outputs (e.g., electricity or refinery products). Facilities can operate to meet local demand, to meet export targets, or other criteria. For the electricity sector, for example, specific plants can be dispatched to meet an annual system load curve.

3.7.3.3 Calibration to base year energy balance

Once detailed data structures have been established and data entered for the base year, the full data set should be calibrated to reflect official base year energy supply totals. When summed across end uses, subsectors, and sectors, the total energy use and required supply, determined from a set of disaggregate "bottom-up" data, will often disagree with official supply totals. Disaggregated data must then be reviewed and adjusted, based upon the analyst's judgment. Large differences generally indicate the need for closer analysis. Small differences can be allocated to the least accurately defined demand categories, such as "other industries", or can be allocated to all energy use categories on a pro-rata basis.

3.7.4 Setting Up Emissions Data

Setting up the calculations by which fuel use estimates are translated into estimates of the emissions of GHG and other pollutants has three elements:

- **Decide which gases to consider.** Energy sector activities release a number of different types of pollutants in varying quantities. The analyst must reduce the large field of pollutants to a number that is manageable for the study. Among GHG, CO_2 is almost always the GHG released from energy sector activities in the largest quantities. Other GHG that can be of importance include methane (CH_4) and N_2O. The analyst may also wish to consider other pollutant emissions, such as SO_x and NO_x.

- **Specify emission factors for each energy activity.** Relevant emission factors should be specified for each appropriate energy conversion, transport, and production activity, and each appropriate energy-consuming activity. If not available from the GHG Inventory, GHG emission factors can be taken from the IPCC and other sources, as listed inTable 3-3.

- **Calibrate base year emissions.** Once connections have been made between energy data and emission factors, the resulting national energy sector emission estimates should be compared and calibrated to the GHG emissions inventory, if available. Differences between the base year emissions estimates from the mitigation study and an official GHG inventory may be due to a number of factors, including differences in energy data, differences in emission factors, or off-line computational errors.

3.8 DEVELOPING A BASELINE SCENARIO

Developing a baseline scenario that portrays social, demographic, and technological development over a 20-40 year or longer time horizon can be one of the most challenging aspects of a mitigation analysis. The levels of projected future baseline emissions shape the amount of reductions required if a specific mitigation scenario target is specified, and the relative impacts and desirability of specific mitigation options. For instance, if many low-cost energy-efficiency improvements are adopted in the baseline scenario, this would yield lower baseline emissions and leave less room for these improvements to have an impact in a mitigation scenario.

Development of a baseline scenario begins with the definition of scenario characteristics (e.g., "business-as-usual"). Changes in exogenous driving variables must then be specified and entered into the model, which is then run to simulate overall energy use and emissions over the time horizon selected. The baseline scenario must be evaluated for reasonableness and consistency, and revised accordingly. Uncertainty in the evolution of the baseline scenario can be reflected through sensitivity analysis of key parameters such as GDP growth.

The procedure will vary somewhat depending on the modeling approach used and the nature of a baseline scenario. In an optimization model, the use of different technologies is to a certain degree decided within the model, dependent on how much one wants to constrain the evolution of the baseline scenario. For example, the analyst might choose to construct a baseline scenario in which the energy supply system closely reflects or extrapolates from published plans. Alternately, the analyst might choose to give the model more flexibility to select a future energy supply system based on specific criteria. If one is using an optimization model, it is necessary to introduce certain constraints if one wishes to force the model toward a solution that approximates a "business-as-usual" future.

The rate of economic growth and changes in domestic energy markets are among the most important assumptions affecting projected baseline emissions. Official government GDP projections may differ from other macroeconomic projections. In terms of domestic energy markets, the removal of energy price subsidies could greatly affect fuel choice and energy efficiency, and thus baseline emissions and the impacts of mitigation options.

A preparatory step in developing a baseline scenario is to assemble available forecasts, projections, or plans. These might include national economic development plans, demographic and economic projections, sector-specific plans (e.g., expansion plans for the iron and steel industry), plans for transport and other infrastructure, studies of trends in energy use (economy-wide, by sector, or by end use), plans for investments in energy facilities (electricity expansion plans, new gas pipelines, etc.), studies of resource availability, and projections of future resource prices. In short, all studies that attempt to look into a country's future—or even the future of a region—may provide useful information for the specification of a baseline scenario. However, it is unlikely that every parameter needed to complete the baseline scenario will be found in national documents, or even that the documents will provide a consistent picture of a country's future. As with much of the modeling process, the judgment of the analyst in making reasonable assumptions and choices is indispensable.

3.8.1 Developing Projections of Energy Demand

In bottom-up approaches, projections of future energy demand are based on two parameters for each subsector or end use considered: a measure of the activity that drives energy demand and a measure of the energy intensity of each activity, expressed in energy units per unit of activity.

Measures of activity include data on household numbers, production of key industrial products, and demand for transport and services. Activity may be measured in aggregate terms at the sectoral level (e.g., total industrial value added, total passenger-km or tonne-km), and by using indicators at the subsector level. These two measures need not be identical. For example, total industrial value added is a common indicator for aggregate activity for the industrial sector, but for specific subsectors, such as steel or cement, one often uses tonnes of production as a measure of activity. In general, physical measures of activity are preferable, but they are not appropriate in all cases (such as in light industry, where there is no aggregate measure of physical production). Measures of activity are discussed in more detail in the chapters on each energy demand sector (Chapters 4-7).

In bottom-up approaches, future values for driving activities are exogenous. In other words, they are based on external estimates or projections rather than being estimated by the model itself. Future values can be drawn from a variety of sources or estimated using various forecasting methods. Estimates of the future levels of activity or equipment ownership will depend on local conditions and the behavioral and functional relationships within each sector.

Projections of the future development of energy intensities in each sub-sector or end use can be expressed in terms of final energy or useful energy.[3] When the choice of end-use options is conducted within the model, however, the energy demand should be given in terms of useful energy to allow the model to select among technologies for meeting the energy requirements.

---

[3] "Final energy" is the energy delivered to end-use devices. "Useful energy" refers to the amount of energy required to meet particular demands for energy services. It is typically estimated by multiplying final energy consumption by the average conversion efficiency of end-use equipment (e.g., of oil-fired boilers). Applying the concept of useful energy is more difficult in the transport sector although one can use the estimated conversion efficiency of generic engine types in various vehicle classes.

Projections should start from the base year values, such that the sum of the product of energy intensity and activity level in each sub-sector add up to total final or useful energy use in the base year. In energy statistics, the data are normally presented in terms of final energy. If the projections are to be given in useful energy units, the statistical data should be converted using estimated base year efficiencies for each end use.

Ideally, the projections of energy intensities should be based in part on historical developments. To the extent statistical data on a disaggregated level are available, they give limited information for economies that have undergone significant structural changes or changes in taxation/subsidies on energy. Even if reliable historical data are available, assumptions on the development of the energy intensities have to be made using careful judgment about the status of existing end-use technologies and future efficiency improvements that should be included in the projections.

It is important to distinguish between improvements included in the exogenous demand projections and improvements that are explicitly included as technology options in the model. For example, future improvements of building insulation standards can either be directly included in the demand projections as an improvement of intensity for space heating, or modeled as "technology options" that can be chosen individually in the assessment, depending on their attractiveness in the different scenarios. The distinction is especially important in optimization models. One can assume that the insulation standards will be implemented (e.g., through use of regulations), or allow the model to choose the implementation. In the latter case, the insulation option will be implemented if its cost is less than the cost of supplying the heat using available options for space heating.

3.8.2 Calculate and Review Results of the Baseline Scenario

Once the assumptions to be used in the baseline scenario have been mapped out, the next step is to enter the projected values for key parameters. When the data set is complete, the model calculates the results of the baseline scenario. The analyst should check results to ensure that they appear reasonable and logical. When reviewing initial demand results, many questions should be raised, such as: Do the growth rates of fuel demand in each sector make sense, based on the input assumptions, past trends, and expected developments? Are there changes in demand patterns that seem unlikely?

When reviewing supply results, the analyst may need to ensure that all fuel demands are being met, either through domestic production or by imports. For the electricity system, the analyst should make sure that an adequate reserve margin is maintained throughout the planning period and that there is a reasonable balance between baseload, intermediate, and peaking resources.[4] If supply resources are insufficient, demands should be curtailed, prices augmented, or new supply resources added. If excess capacity (e.g., unneeded power plants) and/or resources are present, expansion plans should be delayed or curtailed, if possible. (In an optimization model or a system simulation model, these supply-side "checks" are generally performed by the model.)

The final step is to calculate the emissions consequences of the scenario. Here again, the analyst should review results to see if they are reasonable. Are the emissions of the expected order of magnitude? If not, and errors in the data sets have been eliminated as possible causes, some emission factors may be incorrect or missing.

---

[4] The reserve margin is the difference between total installed electric generation capacity and the annual peak demand, divided by the peak demand. In general, reserve margins of 15 to 25% are considered adequate for ensuring reliability of electric service. Larger reserve margins might indicate an electricity generation system that is over-built (and thus more expensive than necessary) or poorly maintained. Baseload resources are power plants that run most of the time to meet a constant level of demand day and night. Peaking plants run less frequently to meet demand only at "peak" periods (e.g., in the morning or evening) when demand is highest.

Once the initial baseline scenario is prepared, it should be reviewed to assess whether it presents a comprehensive and plausible future for the country in light of real-world constraints. Some specific questions might include:

- Can the indicated growth rates in energy demand be sustained over the study period? Is it a reasonable rate, given recent experience in the country and region?
- Is the level of capital investment needed to sustain the indicated levels of industrial growth likely to be forthcoming?
- Will the country be able to afford the bill for fuel imports that is implied by the baseline scenario?
- Will the level of capital investment needed for energy supply system expansion be available given competition for limited financial resources?
- Is the indicated increase in transportation use plausible given current and planned transportation infrastructure?
- Are the emission factors in use appropriate for future technologies?[5]

Answers to these questions might indicate the need for adjustments to the baseline scenario or sensitivity analyses of key parameters.

3.9 DEVELOPING MITIGATION SCENARIOS

The process of developing a mitigation scenario or scenarios involves establishing a scenario objective and combining specific options in an integrated scenario. Integrated scenario analysis is essential for developing accurate and internally consistent estimates of overall cost and emissions impacts since the actual emissions reduction from employing a specific option can depend on the other options included in a scenario. For instance, the level of reduction in GHG emissions associated with an option that saves electricity is dependent on the electricity generation resources whose use would be avoided (e.g., coal, oil, hydro, or a mix). In reality, the type of electricity generation that the efficiency option would avoid will change over time, and, if lower GHG-emitting electricity resources are also introduced in the scenario, the GHG savings of the efficiency option may be reduced. Integrated scenario analyses are intended to capture these and other interactive effects.

Where using an optimization model, the difference in input data for the baseline scenario and the mitigation scenario(s) is typically less than in an accounting model, where the choice of technologies is exogenous to the model. An optimization model chooses from the whole set of available technologies to satisfy the given constraints.

3.9.1 Objectives for Mitigation Scenarios

Several objectives are possible for designing a mitigation scenario. The objective will depend on political and practical considerations. Types of objectives include:

- Emission reduction targets. For example: 12.5% reduction in CO_2 emissions by 2010, and 25% reduction by 2030, from baseline levels. An alternative is to specify reductions from base

---

[5] CO_2 emissions per unit of fuel use (or production) are unlikely to change much as new technologies are introduced. Emission factors for other gases are more likely to be affected by technological changes.

year levels, which avoids making the amount of reduction dependent on the specification of the baseline scenario.

- Identification of options up to a certain cost per tonne of emissions reduction. The emissions reduction given by the resulting technology mix would reflect the level of reduction that could be achieved at a certain marginal cost.

- "No regrets" scenario. This scenario is a common variant of the previous type of objective, where the screening threshold is essentially zero cost per tonne of GHG reduced.

- Specific options or packages of options. Examples of this type of scenario might be a "natural gas" scenario, a "renewable energy" scenario, or a "nuclear" scenario.

Another important aspect of defining the scenario objective is to decide on whether only CO_2 or several GHGs will targeted. Non-CO_2 GHGs, such as CH_4 (methane) or N_2O, can be separately targeted for reduction, or a combined reduction target can be specified using global warming potentials (GWPs).

3.9.2 Developing Integrated Scenario(s)

The manner in which one develops an integrated scenario containing a combination of mitigation technology options differs among energy sector models. In an accounting model such as LEAP, the analyst selects specific options based on their cost of saved carbon and other characteristics, and specifies their penetration. In an optimization or simulation model, the selection of technologies is done by the model according to its internal logic and criteria specified by the user.

It is often desirable to develop several mitigation scenarios, as different combinations of options might have similar total direct costs but differ with respect to criteria such as balance of payments or local environmental well-being. Since many objectives are important to national policy, and modeling approaches cannot capture all costs and benefits, the absolute lowest cost scenario may not be the preferable one.

In addition, the analyst may choose to assemble a range of plausible assumptions for key input variables and test their impacts as sensitivity analyses.

3.9.2.1 Penetration rates of mitigation technologies

A key task is to determine reasonable penetration rates for technology options. Penetration rates denote the speed at which a technology or practice can be adopted ("enter the market"). For large investments, such as power plants or large industrial investments, the penetration rate will generally be expressed in terms of the timing and size of discrete changes or capacity additions. For instance, for technical and infrastructure reasons, there might be a limited amount of wind power that can be installed by a specific future date. Alternatively, the intermittent nature of wind power generation might limit its maximum achievable penetration if no storage capacity is available.

For smaller investments, such as consumer appliances or smaller industrial investments, penetration rates are typically expressed in terms of the percent penetration per year of improved technologies or practices. For instance, an achievable market penetration rate for high-efficiency refrigerators might be 50% of new purchases per year. If 6% of all refrigerators are replaced (or purchased by households who previously had none) each year, the total penetration rate of energy-efficient refrigerators into the total stock of refrigerators would be 3% per year. In addition to annual penetration rates, there might be a maximum achievable penetration level, akin to the limitation on wind power implementation described above, if a certain percentage of the market is unlikely to adopt a option for technical or behavioral reasons.

In general, lower-GHG technologies can be introduced either (a) when existing equipment reaches the end of its economic life and needs replacement; (b) when new equipment is needed to meet growing demands; (c) as a retrofit measure which modifies existing equipment; or (d) by replacing existing equipment before the end of its economic lifetime. As a consequence, several factors can influence penetration rates, including:

Equipment lifetime. For example, a coal plant may last for 30 years, a refrigerator for 15 years, and an incandescent light bulb for 1 year or less. Stock models can be used to track the replacement of "vintages" of technologies and indicate the timing and opportunity for technological improvements. Stock models can be represented in the integrating energy models. Alternately, more detailed stock models may be used for specific technologies.

Technical, infrastructure, or financing limitations. The local availability of products, technologies, skilled labor, and capital can influence how rapidly a mitigation option can be adopted.

The policy instrument used. For example, to accelerate the adoption of energy-efficient refrigerators, either efficiency standards or incentive programs could be used; the two options would likely achieve rather different penetration rates.

In an accounting model, the analyst specifies a particular penetration rate for each option. With an optimization model, the analyst typically specifies maximum (or minimum) penetration rates that restrict the model's selection.

3.9.2.2 Review Results

A model will produce a variety of reports, including summaries of costs, savings, and emissions reductions achieved. The analyst should review the results for consistency and reasonableness. In an accounting model, the energy supply system should be checked to ensure that it is sufficient to meet projected demands without significant excess capacity. Reduced electricity demand will delay the need for new capacity, and the results should be reviewed to ensure that a reasonable reserve margin is maintained but not substantially exceeded.

3.10 COST CURVES FOR GHG ABATEMENT[6]

GHG reduction in the energy system can be achieved through changes at many stages of the system, involving a wide variety of different energy technologies. Fully cost-efficient abatement should consider all available abatement options, at end-use, conversion, and production stages, on an equal footing. A cost-efficiency analysis of GHG abatement in the energy system should therefore be able to compare all these different options in an integrated procedure. A cost curve expressing the costs per unit of emission reduction as a function of quantity of GHG reduced must thus be an aggregate of a number of different technical and structural changes in the energy system, many of which are interdependent. The capability of representing these interdependencies and how the detailed energy system information is mapped into a one-dimensional form differs significantly among the different models.

---

[6] Parts of this section draw on the UNEP Greenhouse Gas Abatement Costing Study, Phase One Report (UNEP, 1992).

In an accounting model, it is implicitly assumed that the economic and GHG emission "value" of each technology can be determined independently because each technology is assumed to represent a marginal adjustment of the total system. However, if fundamental, non-marginal, structural changes of the supply system, such as fuel substitution and the introduction of combined heat and power (CHP) systems, are considered, the economic and GHG emission "values" for the energy demand technologies are not unique and cannot be determined independently of the supply system. Any investment alone can influence the total economy and GHG intensity of the energy system, and technologies must therefore be assigned different economic and GHG emission "values" depending on the system in which they take part.

In integrated models like ENPEP and MARKAL, the supply side responds to even marginal changes on the demand side. However, this integrated approach also makes it difficult to relate emission reductions at each point of the curve to specific single-technology options.

There are several ways of establishing cost curves and which method to chose depend largely on the type of model used. In the following, we discuss three different approaches for constructing cost curves.

3.10.1 Methods for Construction of Cost Curves

- **The partial solution**. In this method, each technology option is evaluated separately with respect to incurred costs and GHG emissions. Results are compared, option by option, for the GHG-reducing option with the baseline. Alternatively, the GHG-reducing option might be compared to a total reference case. Incremental changes in GHG emissions and costs are calculated and ranked according to costs as shown in Figure 2-2. Here option A is chosen as the most valuable option.

  The procedure is little more than a stacking of abatement options. Ranking of the options is done separately for each, ignoring interdependence among the options. The cost curve is built up of partial independent segments. The solution is simple, but theoretically unsatisfactory, and can lead to serious misinterpretations.

- **The retrospective systems approach**. This approach requires the development of a baseline with which to compare the mitigation case. The first step in this approach is a separate ranking of the technology options, using an energy systems framework. This step is identical to the partial approach. In the next step, the most valuable option (option A in Figure 2-2) is included in the model run, and incremental results compared to the reference case are calculated to give the block in the cost curve. Next, the second most valuable project is included in the model and new calculations are performed. Results are compared to the former ones, a new cost curve block is established, and so on.

  The advantage of this approach compared to the partial solution is that the interdependence between the given option and every other previous option on the cost curve is taken into account. The results for option B will thus depend on the introduction of option A, while option C will depend on both options A and B and so on. However, this approach does not include the impacts that the implementation of more expensive options can have on the options already chosen. For example, if option G in Figure 2-2 is a hydro power plant, implementation of this option can change both the CO_2 intensity and cost of the electricity in the energy system. If option E is an option for saving electricity at the end-use level, the implementation of option G will change both the amount of GHG reduced and the costs for option E. This lack of consistency in the retrospective approach is an important drawback, especially in cases where implementation of demand-side options results in significant changes on the supply side. Another drawback is that the retrospective method implies that once an abatement option is included in a scenario, it will be a permanent part of all subsequent scenarios.

Both the partial solution and the retrospective approach result in a stepwise cost curve, and these two methodologies are often applied when establishing cost curves using accounting models.

- **The integrated systems approach**. This approach requires the existence of a well-defined baseline, and an integrated energy system model where the supply side automatically responds to even marginal changes on the demand side in each model run. Both ENPEP and MARKAL facilitate this fully integrated approach. The idea is to determine any point on the cost curve as the least-cost solution for the total energy system, in which all demand and supply system parameters can vary. This approach aims to take into account all the interdependencies within the system, as represented by the energy systems model.

  A least-cost curve can in principle only be established through use of a cost-minimizing optimization model. In an optimization approach, the model itself selects the configuration of the energy system that satisfies different emission targets at a minimum cost. To identify a theoretically consistent curve showing the least cost at any emission reduction level, the model should have the same options available to chose from in each run with increasing emission reduction levels. In simulation models like ENPEP, one can also establish a one-to-one relationship between emission reduction and cost through imposing constraints on emission levels in the model. However, this will not necessarily reflect a strictly least-cost curve.

  The disadvantage of the integrated systems approach is that the interdependencies between supply and demand options make it difficult to identify impacts of single options separately, and a strict ranking of options is not possible since the emission reductions achieved as one moves up the cost curve result from the use of several interlinked options.[7]

The difference can be illustrated by comparing Figure 2-2 with Figure 3-3. Instead of showing the contribution from different technology options step by step (as in Figure 2-2), the interval where the options contribute in the integrated systems approach can be illustrated as shown in Figure 3-3.

As the figure illustrates, emission reductions at each level are achieved through the use of a mix of options, rather than stepwise reduction through single options. Despite the difficulty of singling out specific options at each reduction level, the approach represented in Figure 3-3 is still the most satisfactory since it more closely reflects the real situation in an interlinked energy system.

---

[7] The impacts on costs and/or emissions of introducing single options can be studied in an integrated model through running the model with and without the option available. The difference in costs and/or emissions between the two cases yields the cost/benefit of the specific options in a fully integrated system. For example, the value for the entire energy system of introducing high-efficiency refrigerators can be calculated by comparing the total system costs of stabilizing emissions when the refrigerators are available to the case when they are not. In optimization models, the "shadow prices" give information on the relative value to the energy system of making one unit of a specific technology more available. This makes it possible to compare technologies on the same scale independent on whether they are supply or demand technologies. Ranking of options that are not chosen in the optimal solution can also be done through investigating their so-called "reduce costs," which is the reduction of unit cost that is needed to make the technology competitive in the system. Technologies that appear near to the optimum in a linear programming solution could be attractive in a real world where more complex decision variables are involved.

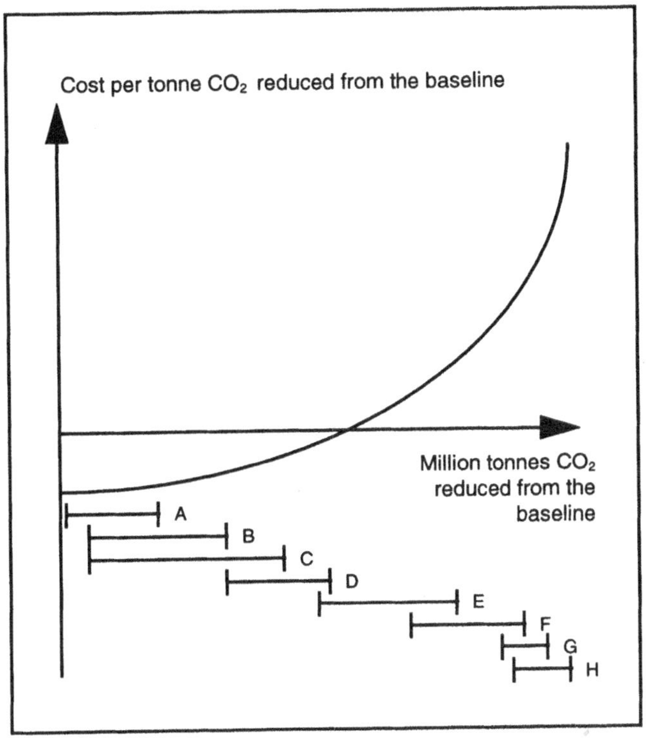

Figure 3-3. CO_2-Reduction Curve Using the Integrated Systems Approach

As mentioned in Chapter 2, it is important to be aware of what the costs presented in the curves include. When derived from engineering models, cost curves usually address direct energy system costs. As such, they contain the same limitations as the models themselves in not incorporating transaction and other hidden costs, which may vary considerably between different options. Moreover, they represent a sectoral notion of costs as opposed to macroeconomic measures such as GDP impacts.

3.11 EVALUATING MACROECONOMIC IMPLICATIONS OF MITIGATION SCENARIOS

The "bottom-up" approach only captures the direct economic impacts of technologies and, if included, the costs of policy instruments—efficiency programs, standards, import tariffs, taxes, and so on—that might be used to achieve adoption of the technology options selected in a mitigation scenario. It is desirable to conduct some assessment of the impacts on the macro-economy in terms of growth in a particular sector, employment, and other indicators. One approach for evaluating some of these economic impacts is to integrate bottom-up results with "top-down", macroeconomic analyses. Capital investment, fuel expenditures, and import and export levels can be used as inputs to a top-down model to assess the macroeconomic impacts of mitigation scenarios.

Bottom-up/top-down integration may be done at different levels of sophistication. At the simplest level, the capital investment, fuel expenditures, and import and export levels required for energy activities may be compared with those for the whole economy. Alternatively, they may be compared with a country's GDP.

Estimating the effects of energy-sector scenarios requires a linkage to a macroeconomic model. The one-sector MACRO component of MARKAL-MACRO provides a way to estimate the impact on GDP of pursuing different scenarios. In a multi-sector framework, the energy sector results may be used as final demand drivers in an input-output (IO) framework. The IO model will yield estimates of direct and indirect incremental changes in output, employment, environmental, and other associated impacts. The estimates may also be used in conjunction with a CGE model (e.g., the LBL-CGE model) to determine the impact on the government budget, for example. Regardless of which approach is used, it is extremely important for the modeler to place the bottom-up results in the context of macroeconomic parameters which are important to policy-makers.

REFERENCES

Buehring, W.A., Hamilton, B.P., Guziel, K.A., Cirillo, R.R., Conzelmann, G., and Koritarov, V. 1994. ENPEP: An Integrated Approach for Modeling Energy Systems. Argonne National Laboratory, Argonne, IL. September 1994.

Eckhaus, R.S. and S. Lahiri. 1994. A General Description of MIMEC (Multisectoral, Intertemporal Model of Emissions Constraints). Massachusetts Institute of Technology, Department of Economics, Cambridge, MA.

Electric Power Research Institute. 1993. Technical Assessment Guide (TAG). EPRI TR-102276-V1R7. Electric Power Research Institute, Palo Alto, California, U.S.A.

Geller, H. 1991. Efficient Electricity Use: A Development Strategy For Brazil. Washington, D.C.: American Council for an Energy-Efficient Economy.

Intergovernmental Panel on Climate Change. 1992. 1992 IPCC Supplement: Scientific Assessment of Climate Change, IPCC Working Group 1, WMO/UNEP.

Johansson, T.B., Kelly, H., Reddy, A.K.N., Williams, R.H., Burnham, L. 1992. Renewable Energy: Sources for Fuels and Electricity. Island Press.

Kram, T. 1993. National Energy Options for Reducing CO_2 Emissions, Volume I: The International Connecti`on. Netherlands Energy Research Foundation (ECN).

La Rovere, E.L. and L.F.L. Legey, J.D.G. Miguez. 1994. Alternative Energy Strategies for Abatement of Carbon Emissions in Brazil: A Cost-Benefit Analysis. *Energy Policy*, November.

Manne, A.S. and R.G. Richels 1992. Buying Greenhouse Insurance: The Economic Costs of Carbon Dioxide Emission Limitations. The MIT Press, Cambridge, MA.

Manne, A.S. and G.A. Goldstein, L.D. Hamilton, J. Lee, W. Marcuse, S.C. Morris, C. Wene. 1992. *MARKAL-MACRO: An Overview.* Brookhaven National Laboratory, Informal Report #BNL-48377, Upton, NY.

Mongia, N., and R. Bhatia, J. Sathaye, P. Mongia. 1991. Cost of Reducing CO_2 Emissions from India: Imperatives for International Transfer of Resources and Technologies. *Energy Policy*, December.

Reddy, A.K.N. 1991. Environmentally Sound Energy Development: A Case Study of Electricity for Karnataka State. Presented at the Regional Conference on Environmental Challenges for Asia-Pacific Energy Systems in the 1990s, Kuala Lumpur, Malaysia. January 14-18.

Sathaye, J., A. Ketoff, L. Schipper, and S. Lele. 1989. "An End-Use Approach to Development of Long-Term Energy Demand Scenarios for Developing Countries," Lawrence Berkeley Laboratory Report LBL-25611, Berkeley, CA.

Schipper, L. and S. Meyers. 1992. Energy Efficiency and Human Activity: Past Trends, Future Prospects. Cambridge University Press, Cambridge, UK.

SEI-B (Stockholm Environment Institute - Boston). 1993. *The Long-range Energy Alternatives Planning System, Overview for LEAP Version 94.0.*

United Nations Environment Programme. 1992. UNEP Greenhouse Gas Abatement Costing Studies, Phase One Report. UNEP Collaboratory Centre on Energy and Environment, Riso National Laboratory, Denmark. (August).

United Nations Environment Programme. 1994a. UNEP Greenhouse Gas Abatement Costing Studies, Phase Two, Main Report. UNEP Collaborating Centre on Energy and Environment, Riso National Laboratory, Denmark. (May).

United Nations Environment Programme. 1994b. UNEP Greenhouse Gas Abatement Costing Studies, Phase Two, Guidelines. UNEP Collaborating Centre on Energy and Environment, Riso National Laboratory, Denmark. (May).

United States Department of Energy. 1993. The IPCC Technology Characterization Inventory, Phase II Report, Office of Environmental Analysis, Office of Domestic and International Energy Policy, Developed for the Energy and Industry Subgroup, Response Strategies Working Group of the Intergovernmental Panel On Climate Change.

United States Department of Energy. 1994. Annual Energy Outlook 1994: With Projections to 2010. (1994). Energy Information Administration, U.S. Department of Energy, Washington, D.C.

UCS *et al.* 1991. America's Energy Choices: Investing in a Strong Economy and a Clean Environment. Union of Concerned Scientists, Alliance to Save Energy, American Council for an Energy-Efficient Economy, Natural Resources Defense Council, Tellus Institute. Union of Concerned Scientists: Cambridge, MA.

World Energy Council. 1992. 1992 Survey of Energy Resources. World Energy Council, London, U.K.

CHAPTER 4
INDUSTRIAL SECTOR

4.1 INTRODUCTION

The opportunities for GHG mitigation in industry are somewhat different from those in other sectors because the greatest increases in the efficiency of energy and materials use often come not from direct efforts to reduce consumption, but rather from pursuing other goals such as improved product quality and lower production costs. Many investments undertaken for non-energy reasons yield energy-efficiency gains as a secondary consequence.

Industry is also different in that firms usually have an incentive to improve efficiency in order to reduce costs and maintain their profitability. The degree to which this incentive applies varies among countries depending on the nature of the economy. In the type of market economy toward which many developing and transition countries appear to be evolving, one can expect that industry will pursue a number of the avâilable opportunities for improving the efficiency of resource use. Shortage of capital is a problem in many cases, but gradual improvement in efficiency is likely as investment takes place and new plants are built. Thus, mitigation strategies in industry can build upon and enhance efforts that may already be underway. Well-designed policies can have an important impact, but the overall economic and financial context is likely to be more significant in encouraging investment in energy efficiency than particular policies explicitly designed to have that effect.

4.2 MITIGATION TECHNOLOGY OPTIONS

Actions by industry that can reduce emissions from energy use include more efficient use of energy, fuel switching, and improved use of materials that reduce energy requirements. In addition, policies that lessen the demand for energy-intensive commodities can reduce industrial energy consumption.

Energy efficiency is the most important category for mitigation analysis. There are many technologies and practices that could enhance industrial energy efficiency. Minor operational changes, such as housekeeping and maintenance, are typically the cheapest, easiest to implement, and least risky, but usually yield the smallest energy and cost savings. Production equipment changes and energy conservation add-on technologies involve larger investments, typically thousands to millions of dollars, and may or may not be justified by reduced energy costs alone. Major process changes often require building a new facility and are usually justified only by strategic market development concerns.

Industry consists of a vast array of activities involving thousands of different processes that are often site-specific in design. Whereas the buildings and transportation sectors can utilize a limited number of energy conservation measures that may be widely applied, the industrial sector requires more of a focus on options in specific industries. Generic technologies that cut across industries represent only part of the full range of opportunities, and even these generic technologies are typically customized for particular applications.

Because of the nature of decision-making in the industrial sector, it is useful to consider two basic classes of technical options: (1) actions for which energy-cost savings are the dominant criteria, and (2) actions for which broader criteria such as overall production cost and product quality are the dominant criteria. One could call the former "energy-cost-sensitive options," and the latter "non-energy-cost-sensitive options."

4.2.1 Energy-Cost-Sensitive Options

Energy-cost-sensitive options include low- to medium-cost improvements to the energy efficiency of existing capital stock, production and use of more energy-efficient equipment, and fuel switching. Many energy-efficiency options are applicable across industries. These generic options often have benefits in addition to energy-cost savings that may be significant to their adoption.

4.2.1.1 Measures for existing processes

Some examples of energy-cost-sensitive measures that can enhance energy efficiency in existing plants are described below.[1]

- **Housekeeping, equipment maintenance, and energy accounting.** Good housekeeping includes activities such as carrying out inspections to encourage conservation; scheduling energy-intensive activities; turning off equipment when not in use; installing and using energy-monitoring equipment; wrapping tanks and pipes with insulation; and repairing leaks. Regular equipment maintenance can prevent the loss of efficiency that can occur over time. Energy accounting systems can be used to help motivate energy-conservation activities.

- **Energy management systems** can be used to systematically turn off or turn down process equipment, lights, and fans.

- **Motor drive system improvements** include use of high-efficiency motors, improved motor rewinding, power conditioning, drive control (especially with adjustable-speed drives), and use of more efficient associated equipment (pumps, fans, compressors).

- **Improved steam production and management** includes use of economizers and other heat recovery systems, attention to steam distribution systems, and use of more efficient boilers.

- **Industrial cogeneration** allows the substitution of waste heat from electricity generation for steam that would otherwise be raised in a boiler using fuel (this describes its most common application, the "topping cycle"). Cogeneration is an important option for industries with large process heat requirements such as pulp and paper, chemicals, and food processing. It is particularly attractive where there are on-site energy sources that are not being utilized.

- **Heat recovery** may involve transferring heat from high-temperature waste heat sources to more useful media such as steam or raising the temperature of low-temperature streams so they can be useful as heat sources.

---

[1]Examples of successful implementation of energy-efficiency measures in a developing country are described in a recent report from India (NPC, 1993).

4.2.1.2 Measures for new equipment

Adoption of more energy-efficient new equipment is an option for devices widely used across industries, such as electric motors, pumps, fans, compressors, and boilers. It is an especially important option in countries where industry is growing rapidly and the manufactured equipment is outdated.

4.2.1.3 Fuel switching

Opportunities for switching to lower-carbon fuels vary among countries depending on the available resources. The most important options in a mitigation analysis are switching to natural gas or renewable energy sources (e.g., wood from managed plantations for boilers or solar thermal energy for low-temperature process heat demands).

4.2.2 Non-Energy-Cost-Sensitive Options

Non-energy-cost-sensitive options include major modifications to existing production capacity and addition of new production capacity that incorporates state-of-the-art technology. These options are usually specific to particular industries. Improved use of materials (through recycling and process yield improvements) can be considered in this category.

- **The nature of major modifications** to existing plants varies by industry. Some examples are improvements to electric arc furnaces and revamping open-hearth furnaces where they are still viable (steel), installing an improved aluminum smelter, improved ethylene cracking, and conversion from semi-dry to dry process or installation of pre-calcination (cement). Across industries, automated process controls based on new sensor technologies can improve product uniformity and quality and reduce waste from the product stream.

- **Installation of new production capacity.** New capacity is more productive and usually more energy-efficient than existing capacity. The degree of difference depends on the nature of the new technology—which may be a slightly improved version of existing technology or an entirely new technique, the design of new systems, and the age and condition of existing plants. New capacity allows for application of improved technology, design, and process control. Major energy savings are often possible with process integration, which involves designing processes so that the number of heating and cooling steps are minimized. Proper sizing of equipment and matching of components can yield significant savings, as can use of controls and sensors.

  Promoting adoption of state-of-the-art technology is an important option over the 10-20 year time horizon during which capital stock is replaced and updated. Near-term adoption of advanced technology is probably not a realistic option for most developing and transition countries, but may be viable if one is considering long-term potential.[2]

---

[2] State-of-the-art technology refers to the best available technology with demonstrated technical feasibility in actual production environments, such as in a commercial or large-scale plant. Advanced technology refers to technologies that are under development or have been seriously considered in concept and are expected to have an impact on the industry over the next 10 to 25 years. Companies will usually want to introduce advanced technologies in industrialized countries where the necessary engineering support is available.

- **More efficient use of materials.** Recycling scrap, whether from downstream fabricators or post-consumer wastes, bypasses the most energy-intensive steps of manufacturing—the conversion of ores and feedstocks into basic materials. The largest energy savings are available in aluminum production. Smaller, yet significant, savings are possible for steel and glass. Process yield improvements and quality controls save energy by reducing the amount of material that must be processed to provide the desired output. Continuous casting in the steel industry is an example of a yield-improving technology that saves large amounts of energy. Automated process controls and sensors are an integral part of most yield-improving technologies.

4.2.3 Screening Options

Those industries having the largest emissions, or prospects for major growth, are most important to target in a mitigation assessment. Screening of technology options within a given industry can use the criteria listed in Table 2-2 in Chapter 2. Data availability and the resources available to collect data are important to consider.

4.3 INPUTS FOR INDUSTRIAL SECTOR ANALYSIS

4.3.1 Data on Production and Energy Consumption in the Base Year

A common approach is to separately consider each of the major energy-intensive industries and place all other industry in a single sub-sector. Along with the traditional energy-intensive industries (steel, cement, basic chemicals, etc.), one may also consider industries that are locally important (e.g., food processing or textiles). Units of physical production (such as tonnes) are preferable. For industries with a heterogenous product mix, economic measures such as value added can substitute for physical output.

For certain industries, one may wish to disaggregate total production by product type. For example, one can disaggregate total steel production into carbon, stainless, and alloy steel. This kind of disaggregation may be difficult to estimate, but it allows for a more precise selection of technology options.

4.3.2 Forecast of Production in Each Sub-sector

Forecasts of production should draw on the assumed values for growth of GDP and industrial value added. For energy-intensive industries, projections of physical production can be based on historical relationships between production and GDP, with consideration of export/import possibilities. For basic materials such as steel, paper, and cement, one should also consider whether growth in per capita domestic consumption is on a rising or slowing path.[3] Materials substitution, such as use of plastics and aluminum instead of steel in automobiles, and use of a lower quantity of materials in products, such as in thinner, higher-quality steel, are important to long-run trends in demand for energy-intensive products (Williams *et al.* 1987), as are changes in export strategies. For other industries, the projection may be in terms of value added.

---

[3] In the U.S., per capita consumption of steel and cement ceased its historical increase in the 1970s and began a steady decline due to structural shifts in the economy and changes in manufacturing methods (Williams *et al.*, 1987). Nearly all developing countries are still on a rising path, but the rate of increase is likely to slow as an economy becomes more developed. The transition countries, on the other hand, have historically had relatively high levels of per capita consumption of many basic materials, and for these countries a decline in consumption is likely.

It is especially important to understand the future of the major energy-intensive industries in a country. This involves: a) understanding the current industry as to product quality, modernity of process technology, energy intensity, etc.; b) projecting domestic consumption using international comparisons, income levels, stage of technical sophistication of the general economy; and c) estimating unit costs for future production which could serve as a guide to estimating the role of imports or exports.

4.3.3 Technology Data

Analysis of the industrial sector often involves characterizing production technology options that can contribute to meeting each industry's future output. Such analysis works best for industries that produce a relatively homogenous mix of products (such as steel, cement, pulp and paper, oil refining). For industries that produce a more heterogenous mix of products, such as the chemical industry, one can focus on a few energy-intensive intermediate products (such as ammonia or ethylene). For light industries, which typically account for a large fraction of industrial value added but a small share of energy use, it is usually difficult to consider specific processes. Instead, one can analyze the application of generic measures in these industries.

The two general classes of technology options are (1) investments in existing plants, and (2) introduction of new plants. In each case, one can characterize several options that have different levels of investment and energy use, as well as different environmental implications. Several types of retrofits may be considered for improving efficiency in existing plants. New production capacity may employ various readily available technologies or state-of-the-art technology.

For each option, key criteria are:

- the overall production cost per unit of output (including operating as well as annualized investment cost);
- the capital requirement per unit output; and
- the energy intensity for different fuels and electricity.

Other characteristics that are also important to consider include foreign exchange requirements, employment impacts, and environmental impacts (emissions of air pollutants such as SO_2, NO_x, and particulates; discharge of water pollutants).

An example of characterization of options from an analysis of the aluminum industry in India (Mongia *et al.*, 1994) is shown in Table 4-1. The modernized Soderberg technology option has a lower unit production cost than the Soderberg retrofit, but the latter has much lower emissions and also a lower unit investment.[4] For existing plants using the "pre-baked" method, improved anode baking reduces both the production cost and emissions. Installation of improved smelters requires considerably more investment (all of it requiring foreign exchange), and increases the production cost. A new plant employing improved pre-baked technology, which is the main technology relevant to India, has somewhat lower emissions than the improved existing plant, but has a much higher production cost. Note that different technology options affect the intensities of fuel oil, coal, and electricity in dissimilar ways. Thus, the net impact on emissions depends on the fuel mix for electricity supply.

---

[4] The calculation of production cost incorporates energy prices faced by Indian aluminum producers, which are subsidized. In an analysis from a societal perspective, one should use actual long-run marginal energy costs. The calculation of emissions reflects the Indian electricity supply mix, which is weighted toward coal. Thus, measures that reduce electricity use have a strong impact on emissions.

Table 4-1. Production Technology Options, Aluminum Industry, India

Criteria	Technology Options							
	Soderberg Existing	Soderberg Modernised	Soderberg Retrofit	Prebaked Existing	Prebaked Improved	Prebaked Smelter	Prebaked New	Import
Production Cost (Rs/tonne)	11.61	9.75	10.17	11.45	10.35	11.67	14.19	13.14
Investment (Rs/tonne)	0.00	6.95	1.25	0.00	2.41	16.92	34.40	0.0
Foreign Exchange (Rs/tonne)	0.00	3.37	0.00	0.00	0.00	16.92	34.40	0.0
Fuel Oil (TOE/tonne)	0.20	0.17	0.20	0.17	0.12	0.16	0.16	0.0
Coal TOE/tonne)	1.72	1.60	0.86	1.43	1.05	1.05	1.01	0.0
Electricity (Mwh/tonne)	16.90	15.80	16.20	17.00	16.60	14.50	13.70	0.0
CO_2 Emissions (tonne/tonne)	1.88	1.72	1.02	1.57	1.14	1.14	1.01	0.0

Note: Rs = rupees
Source: Mongia *et al.*, 1994.

For industries that use a variety of processes to produce a heterogenous mix of products, the method described above must be modified. In these cases, one can characterize generic measures such as motor system improvements or housekeeping. Options affecting electricity consumption are generally the most important for light industry.

In addition to the criteria discussed above, the product quality associated with technologies should be considered. A new plant will often have higher production cost per unit of output than an improved existing plant, but its products are usually of consistently higher quality, which means they can command a higher price in the domestic or export market.

Data on the characteristics of different technologies can often be obtained from in-country experts on the particular industry. One can use data from other countries if the technology is essentially the same as would be implemented in the study country, which is the case for certain types of products. For example, studies of a number of industries in the U.S. provide data on mid-1980s state-of-the-art technology and describe advanced technologies likely to be commercialized (Energetics, 1988). For options affecting existing plants, the simplest approach is to consider a typical plant for the industry. If there are only a few plants in the country, one could consider each separately.

4.4 DEVELOPING INDUSTRIAL-SECTOR SCENARIOS

This section describes two approaches that an analyst might use in developing industrial-sector scenarios. The exact method used to construct scenarios will depend on the energy sector model being used. The simple examples given below are more illustrative of the approach used with an accounting model, but they illustrate the basic steps and types of choices that must be made (either by the analyst or internally within the model) to develop scenarios.

4.4.1 Scenarios for Specific Industries: A Simple Accounting Framework

A study team may choose not to conduct a detailed analysis of specific production technology options. A relatively simple method for describing technology options in industries such as steel, aluminum, cement, chlorine, ammonia, and petroleum refining involves comparison with international energy intensities, including predicted state-of-the-art intensities. One needs to have a production forecast for the industry, and at least qualitative information on product mix (in industries like steel and petroleum refining where the issue is significant) and the type of production process used in existing facilities. For heterogenous industries, especially industrial chemicals, the fractional reductions in energy intensity can be based on those estimated for related homogeneous industries.

In the following, cement is used as an example. Table 4-2 shows current average fuel and electricity intensities for a hypothetical country and potential intensities that could be achieved in the future through either retrofit of existing plants or additions of new plants.

Table 4-2. Energy Intensity of Cement-Making Technology Options (Example)

	Current Average, Country X	Current Average, OECD	Retrofit	1990 State-of-the-Art	2010 State-of-the-Art
Fuel (GJ/tonne clinker)	5.0	4.2	3.8	3.15	3.0
Fuel (GJ/tonne cement)	4.5[a]	3.80[a]	3.42[a]	2.84[a]	2.40[b]
Electricity (kWh/ tonne cement)	120	155	150	145	130

Source: Holderbank, 1993.

[a]) clinker/cement = 0.90.
[b]) clinker/cement = 0.80. This requires additional grinding of the steel slag or power plant fly ash used as additive.

We assume that Country X produced 10 million tonnes of cement in the base year of 1990 and that 20 million tonnes is the forecast for 2010. Base case and mitigation scenarios can be created using the categories in Table 4-2. For each scenario, one projects the amount of future production coming from each type of technology. This would typically require some consultation with industry experts and an exercise of judgment.

In the example shown in Table 4-3, half of the production in 2010 continues to come from plants existing in 1990. In the Base Scenario, no production comes from retrofit plants, while in the Mitigation Scenario half of the plants are retrofit. As for new cement plants, in the Base Scenario half of the new production comes from plants using 1990 average technology in the OECD countries, while half comes from plants using 1990 state-of-the-art technology. In the Mitigation Scenario, an effort is made to utilize better technology in new plants. One might also include early retirement of older plants in a mitigation scenario, in which case more of the production in 2010 would come from new plants.

In using this method, it is important to qualitatively justify the more important assumptions and to discuss important technical issues, such as the fraction of additives in the cement, and any fuel switching or use of supplementary fuels (not considered in this example).

A baseline scenario should reflect the fact that many of the technical options that improve the efficiency of energy or materials use will be adopted by industry as it responds to market conditions. The extent of adoption will depend on the incentives, constraints, and regulations faced by industrial managers.

Table 4-3. Cement Industry Scenarios, 2010 (Example)
(Production in million tonnes, except as stated)

Technology Category	Base Scenario	Mitigation Scenario
Existing Plants:		
Current	10	5
Retrofit	0	5
New Plants:		
OECD Average	5	0
1990 State-of-the-Art	5	5
2010 State-of-the-Art	0	5
Total Production	20	20
Total Energy Consumption		
Fuel (million GJ)	78	66
Electricity (billion kWh)	2.7	2.7

4.4.2 Scenarios for Specific Types of Equipment

The method described in this section allows one to estimate the potential impact of policies designed to improve specific types of equipment widely used across industries. It is most easily applied for new equipment. The general approach given below can be incorporated in each of the bottom-up models described in Chapter 3, but the method of selecting technologies for baseline and mitigation scenarios varies among them.

The method accounts for retirement and introduction of existing and new equipment in a simple manner. It requires data on end-use equipment stocks, which are usually not readily available for most types of industrial equipment. If a study team wants to analyze a mitigation option associated with a particular type of equipment (e.g., motors), it may try to collect the necessary data to use this method. The simplest approach is to analyze equipment at the aggregate industry level since separately analyzing equipment within each industry is much more data-intensive.

The method requires projection of the number and characteristics of different types of equipment in the target year. The assumptions and parameters used should be consistent with the industry-wide scenarios. Depending on the data that are available, one may consider a few or many types/sizes of equipment. Considering several sizes is desirable, as the opportunities for efficiency improvement typically vary with size. The steps involved are described further below. The option considered in this simple example is the adoption of energy-efficient new motors instead of standard motors. A similar approach could be used for other types of new equipment.

A comparable approach may also be used to analyze options that would affect existing and new motor systems. In this case, two general options could be considered: (1) retrofit improvement to existing systems, and (2) improved design and use of high-efficiency components in new systems. This analysis can be difficult, however, as information on the characteristics of these systems in a country is hard to obtain, and drawing generalizations as to costs and energy savings impacts is problematic.[5]

Step 1. Characterize the base-year stock of motors. This requires data on the number of motors in each type/size class considered and the average utilization (hours/year) and efficiency (%) in each class. Using average size (kW or hp), utilization, and efficiency in each class, one can calculate average electricity intensity for each class of motor. Stock characterization generally requires survey data.

Step 2. Estimate number of motors in each class in the target year. Estimated growth in the number of motors can be based on projected growth in industrial value added. One should also consider structural trends in the industrial sector that may affect the future distribution of motor types and sizes (refer to the industry-wide scenarios).

Step 3. Estimate the number of base-year (survivors) and new motors in target year. This requires a retirement rate for base-year motors, which should be made in consultation with industry experts. Motors in developing countries are often rewound rather than replaced with a new motor.[6] The number of new motors in the target year is simply the difference between the total number of motors and the number of survivors.

Step 4. Characterize the average energy consumption and cost of new standard and high-efficiency motors; estimate CCE. The best source of data on new motor efficiency is test data for motors available in the study country. Efficiencies reported by motor manufacturers may also be used. Costs can be based on actual motor prices, net of taxes. If high-efficiency motors are not produced locally for a particular type/size, they may be available as an import. The cost used in the analysis should be net of import duties. Estimation of the average energy consumption requires data on typical utilization (hours/year) for each type/size of motor. Estimation of motor lifetime should consider operating conditions in the country. Such data may come from a survey of expert judgement.

Steps 5 and 6. Construct baseline and mitigation scenarios. In the baseline scenario, a "frozen efficiency" case would use 100% standard new motors, while a "likely trends" case would use some mix of standard and high-efficiency new motors. A "technical potential" migitation scenario would incorporate high-efficiency motors in all situations where they are judged to be applicable.[7]

4.5 MITIGATION POLICIES

A range of barriers in developing and transitional countries causes the rate of adoption of efficient technologies and practices to be less than it could be. To encourage the adoption of efficiency improvement and other mitigation measures by industry, a variety of policy options are available.

---

[5] Nadel *et al.* (1991) provides a useful discussion of technical (and policy options) for energy-efficient motors and motor systems.

[6] One could also analyze the option of encouraging purchase of an energy-efficient new motor rather than rewinding a failed motor. Rewinding often results in a loss in efficiency.

[7] For some specialized applications, high-efficiency motors are not commonly made. For applications with limited operating hours, they may not be cost-effective. For some applications, multi-speed motors are more appropriate.

4.5.1 Policy Strategies[8]

Crafting policies to enhance energy efficiency is more challenging for industry than for other sectors of the economy. The greater difficulty arises from the diversity of industrial energy use and from the interconnections between energy and production costs, product quality, environmental compliance, and other sensitive business factors. Several points are clear, however. First, energy efficiency is best promoted through policies that: (1) increase investment in industrial plants, and (2) focus that investment in a manner that encourages adoption of efficient technologies and production methods. Second, the energy conservation and efficiency activities and investments should be consistent with sound business strategy. Energy taxes or mandated investments that are too costly can put domestic companies at a competitive disadvantage.

Increasing investment in industrial plants depends on a healthy business and financial climate. A business environment that includes both economic growth and competition tends to promote investment. Without market growth, corporations may have neither the resources nor the incentive to invest. Without competition, companies are under little pressure to invest. Competition that is vigorous but fair signals to companies that being profitable depends on being efficient. Competition has been an important force behind technology innovation and cost reduction in industrial economies, but progress in industrial energy conservation has also been made in economies where the state plays a strong role, such as China (Levine and Liu, 1992). A financial environment that includes low capital costs and a long-term outlook will encourage industrial investment.

Efficient technologies that are both cost-effective and reliable must be available at the time of investment, and they must be given adequate consideration in investment decisions. Investments can be focused to promote adoption of efficient technologies and practices through information programs, financial incentives, regulations, and technology research, development, and demonstration (RD&D). Policy options in each of these categories are listed in Table 4-4 and described briefly below. They vary widely in their potential energy saving impacts and their costs to the government, businesses, and consumers. To illustrate the range of effects, the specific options are sorted into three distinct levels, in order of increasing government involvement and energy savings. The basic level includes relatively low-cost, simple policy options. The moderate level includes options that are more ambitious and in many cases would require legislation and increased government spending. The aggressive level includes options that are quite ambitious, would require legislation, or an increased government role in energy regulation.

Information Programs. The general lack of concern afforded energy in many corporations is a major barrier to investment in energy-efficiency improvements. This problem can be addressed through policies that raise the profile of energy efficiency as a national and corporate goal. The government could assist by providing technical assistance (audits, training), supporting education and advertising programs, supporting independent equipment testing/rating centers, and establishing equipment-labeling requirements.

Financial Incentives. Financial measures can be used to alter investment patterns in order to promote the various technical objectives. These policy instruments include loan assistance, tax credits and other revisions to the corporate income tax code, and taxation of energy products or emissions. Reducing import duties on energy-efficient equipment (or on components that would get used in manufacturing such equipment) is also an option.

Regulations. Regulations are the most direct method of changing industrial behavior. Among the most viable options for influencing industry's use of energy are equipment efficiency standards, reporting and targeting requirements, and regulation of utilities to encourage industrial demand-side management

---

[8] This section is adapted from OTA (1993).

programs and purchase of cogenerated electricity. By excluding substandard equipment from the market, equipment efficiency standards can have a large impact in a short time. They can also help to lower the price of higher efficiency equipment by increasing the size of its markets. Voluntary agreements with equipment manufacturers are a possible alternative to mandatory regulations.

Research, Development, and Demonstration. Continuous improvement in energy efficiency requires a constant flow of advanced, commercially available technologies, which in turn requires a sustained RD&D effort. Such efforts in developing and transitional countries can reduce dependence on the industrial countries for efficient technologies and also adapt technologies to local conditions. RD&D should stress technologies and processes that combine energy efficiency with more prominent corporate goals such as product quality, labor productivity, and compliance with environmental regulations.

Some of the above policies are best applied in combination. For example, loan assistance or other financial incentives can be combined with an auditing program. Equipment testing is a necessary precursor to standards.

Implementing agents for industrial energy efficiency programs may be government agencies, electric utilities, or organizations that have a specific mandate to promote energy conservation. These actors may also work together to further policy goals.

4.5.2 Options for Procuring Best Technology for New Plants

Some of the policy options discussed above will encourage industrial managers to seek the best available technology when building a new plant, but government can help promote good practices that can result in lower costs and higher chances for trouble-free operation. (In some cases, the government may be directly involved in the planning for new plants.)

For many commodity-type products, new plants often use standard designs which have been built before, and it costs far more to engineer and procure equipment for a plant that differs very much in either scale or technology from the most recently built similar plants. Companies should get competitive bids from potential technology suppliers, which encourages them to quote lower profit margins and put their best people on the project. They should also give clear signals to the bidders on the criteria for selection (e.g., the degree of undemonstrated technology that will be accepted).

4.5.3 Estimating What Policies Can Achieve

Much of the long-run improvement in energy efficiency for a given product is driven by technological progress that tends to reduce all inputs. Such progress is autonomous of changes in energy price and relatively insensitive to policy, apart from general policies that promote investment and R&D (Ross and Steinmeyer, 1990). Support for public and joint public/private RD&D is also important for indigenous technology development or adaptation, especially in the long run. While such policies are quite important for industrial development, they are not likely to be considered as mitigation policies.

For options that are more sensitive to energy costs, the policies discussed above can accelerate adoption of measures that industry might do on its own eventually or induce actions that would be unlikely to occur within a given time frame. Of these, the impact of regulations, especially minimum efficiency standards for equipment, is the least difficult to characterize. For other types of policies, the potential impacts are difficult to assess with accuracy. In these cases, one should make conservative assumptions, specifying clearly what the assumptions are.

Table 4–4. Policy Options for Improving Industrial Energy Efficiency

		Conservation of fossil fuels	Conservation of electricity	Cogeneration	Fuel switching
Information programs					
Technical assistance	Provide government-supported auditing and diagnostic assistance to small- and medium-sized industrial plants.	x	x	x	x
	Develop training and certification programs for energy managers and auditors.	x	x	x	—
	Assist electric utilities with industrial DSM program design.	—	x	—	x
Education and advertising	Promote energy-conserving technologies through workshops and technical literature.	x	x	x	x
	Establish public recognition program for corporate commitments to efficiency.	x	x	—	—
Equipment labeling	Implement energy-efficiency labeling program for generic industrial equipment such as motors, pumps, fans, compressors, and boilers.	x	x	—	—
Financial Incentives					
Loan assistance	Guarantee loans and/or subsidize interest rates for energy-efficiency investments.	x	x	x	x
Investor income tax	Grant tax-free status to dividends earned on provisions bonds used to finance energy-efficiency investments.	x	x	x	x

Energy and carbon taxes	Research effects of taxation on industrial competitiveness and viability of methods for mitigating competitive harm.	x	x	x	x
	Tax fossil fuel use • fuel-specific (e.g., petroleum). • based on heat or carbon content of fuels.	x	x	x	x
	Tax imports of petroleum.	x	—	—	x
	Tax CO_2 emissions	x	—	x	x
Corporate income tax provisions	Grant tax credits and/or accelerated depreciation for efficiency investments.	x	x	x	x
Regulations					
Efficiency standards	Study effectiveness of implementing energy-efficiency standards for generic industrial equipment such as motors, pumps, fans, compressors, and boilers.	x	x	—	—
Utility oversight	Provide incentives to encourage utilities to adopt industrial DSM programs.	x	—	—	—
	Relax electricity-transmission rules to allow sales of cogenerated power to the grid or directly to other industrial plants.	—	—	x	—
Reporting and targeting	Establish efficiency reporting and targeting programs.	x	x	—	—
Research, development and demonstration	Expand technology demonstration efforts.	x	x	x	x
	Expand efforts to improve energy efficiency of industrial equipment and processes.	x	x	—	x

SOURCE: Based on Office of Technology Assessment, 1993.

REFERENCES

EEA (1982). ISTUM-2 Industrial Sector Technology Use Model. Energy and Environmental Analysis, Inc, Arlington, VA.

Energetics Inc. (1988). The U.S. Cement Industry: An Energy Perspective. Report to the U.S. Department of Energy. Columbia, MD.

—. (1988). The U.S. Glass Industry: An Energy Perspective. Report to the U.S. Department of Energy. Columbia, MD.

—. (1988). The U.S. Pulp and Paper Industry: An Energy Perspective. Report to the U.S. Department of Energy. Columbia, MD.

—. (1988). The U.S. Steel Industry: An Energy Perspective. Report to the U.S. Department of Energy. Columbia, MD.

—. (1988). The U.S. Textile Industry: An Energy Perspective. Report to the U.S. Department of Energy. Columbia, MD.

Holderbank Consulting Ltd. (1993). Present and Future Use of Energy in the Cement Industry in Canada. Report to Energy, Mines and Resources Canada.

Jaccard, M., J. Nyboer, and A. Fogwill (1993). "How Big is the Electricity Conservation Potential in Industry?" *The Energy Journal*, Vol. 14, No. 2.

Levine, M. and X. Liu (1992). Energy Conservation Programs in the Peoples Republic of China. Lawrence Berkeley Laboratory Report LBL-29211. Berkeley, CA.

Mongia, N., J. Sathaye and P. Mongia (1994). "Energy Use and Carbon Implications in India. "Focus on Industry," *Energy Policy* Vol. 22, No. 11 (September).

Nadel, S., *et al.* (1991). Energy-Efficient Motor Systems. American Council for an Energy-Efficient Economy, Washington, DC.

NPC (National Productivity Council) (1993). Energy Conservation: The Indian Experience. Ministry of Energy, New Delhi.

OTA (Office of Technology Assessment) (1993). Industrial Energy Efficiency. U.S. Congress, Washington, DC.

Ross, M. and D. Steinmeyer (1990). "Energy for Industry." *Scientific American*, Vol. 263, No. 3 September 1990.

Williams, R.H., E.D. Larson, and M.H. Ross (1987). "Materials, affluence, and industrial energy use," *Annual Review of Energy*, Vol. 12:99-144.

CHAPTER 5
RESIDENTIAL AND COMMERCIAL SECTORS

5.1 INTRODUCTION

Energy use in the residential and commercial sectors primarily results from activities that take place in homes and other buildings. Although the two sectors consist of different classes of buildings having different types of equipment, they are both affected by the same general kinds of mitigation options. A similar analytical approach may be used for both sectors, but its application must take into consideration the different features and dynamics of each sector.

An important feature of the residential sector in developing countries is the growth that is taking place in the demand for energy services. In the industrialized countries, demands for many energy services have become saturated. Growth in population will increase total service demand, but usage per household will not increase dramatically even if incomes rise. For example, the number of refrigerators per household and the amount of volume per household cooled by refrigerators in U.S. residences is unlikely to grow much over the next twenty years. In developing nations, where many people are just now beginning to acquire refrigerators, the demand for the service delivered by refrigerators is increasing rapidly.

5.2 MITIGATION OPTIONS

Mitigation technology options for the residential and commercial sectors may be grouped into four broad categories: (1) efficiency improvements on existing and new building shells, (2) efficiency improvements on new equipment, (3) efficiency improvements on existing equipment, and (4) switching to energy sources/equipment with lower GHG emissions. Within each of these categories, there are a variety of technologies and practices that may be applied to reduce energy use or switch to energy sources with lower GHG emissions. For an overview of energy-efficient technologies for buildings, see Koomey *et al.* (1994) or OTA (1992).

It is important to recognize that individual technologies in buildings often interact. For example, measures that reduce energy use for lighting in commercial buildings also reduce air conditioning requirements, but may increase space heating requirements, depending on the climate. The interaction among measures also means that the energy savings associated with one technology cannot be assessed in isolation from other technologies with which it interacts.

5.2.1 Efficiency Improvements for Existing and New Building Shells

A variety of measures are available for making both existing and new building shells more energy-efficient. These include use of adequate levels of insulation for ceilings, walls, and floors; insulating glazing for windows; control of air infiltration; shading devices or solar control glazing to reduce unwanted solar heat gains through windows; and high-albedo materials, coatings, and paints to reflect shortwave solar radiation and thereby keep building surfaces cooler.[1]

Such measures may affect energy consumption for space heating and/or cooling. Measures can generally be applied more easily and at lower cost in design and construction of new buildings, but

---

[1] Use of high-albedo materials, coatings, and paints can also help keep urban surfaces cooler and thereby keep the urban air temperature from rising (the so-called "urban heat island effect"). Along with tree planting, this measure can reduce energy use for cooling. The size of the impact depends on the local climate and the nature of the urban environment.

considerable reduction in energy use can also be attained via retrofit of existing buildings. The cost-effectiveness of different types of improvements depends heavily on the climate in which a building is located and whether energy is used for both heating and cooling or for only one of the two.

A companion strategy for reducing the need for mechanical heating and cooling is to make effective use of natural heating and cooling. Design techniques include passive solar design, which seeks to utilize energy from the sun by proper placement of windows and incorporation of thermal storage; use of natural ventilation for cooling, such as practiced in much traditional architecture; and planting of trees to block the sun or wind.

5.2.2 Efficiency Improvements for New Equipment

Efficiency improvements on new equipment include both incremental changes to a type of equipment and use of different technology that is more efficient than the technology which it replaces. Examples of the former are increasing insulation in refrigerators and using a more efficient compressor in air conditioners. Various design measures can be applied to any type of equipment to improve its energy efficiency. In the residential sector, the most significant types of equipment are refrigerators, water heaters, and depending on the climate, space heaters and air conditioners.

Cookstoves using biofuels are also important energy users in many developing countries, and various improved designs are available either commercially or as prototypes. The magnitude of net CO_2 emissions from biofuels consumption (i.e., the amount that is harvested on a non-sustainable basis) in any given country is uncertain. For this reason, the IPCC Draft Guidelines for National Greenhouse Gas Inventories requires that net CO_2 emissions from burning of biomass fuels be treated as zero (IPCC/OECD, 1994). Thus, measures that improve the efficiency of biomass use do not affect CO_2 emissions from the residential sector. Rather, they increase biomass stocks. Therefore, options to increase the efficiency of biomass use are discussed under Forestry.

Combustion of biofuels also results in emissions of other GHGs, such as methane, CO, N_2O, and NO_x. The global warming potential of these products of incomplete combustion can be considerable, depending on the time scale chosen and whether indirect warming effects are included. If possible, these gases should be accounted for in a GHG inventory. Measures affecting the efficiency of biofuels use will often result in reduced emissions of these gases.

The full cost-benefit analysis of biomass end-use efficiency options should include the reduced emissions of non-CO_2 gases, the gain in biomass stocks (which should be accounted for under Forestry), as well as the economic and health benefits to the end users.

Examples of use of different technology that is more efficient include compact fluorescent lamps, heat-pump space or water heaters, electronic fluorescent ballasts, and high-pressure sodium lamps for outdoor lighting. All of the above technologies are commercially available, but may not be readily available in all countries.

5.2.3 Efficiency Improvements for Existing Equipment

Efficiency improvements on existing equipment include both physical improvements to equipment to enhance its performance and improved utilization of equipment. Examples of the former are heating equipment tune-up, insulation of heat distribution systems, and use of adjustable-speed drives on motor-driven equipment. Examples of improved equipment utilization include use of energy management systems, which are computerized control systems that can be programmed to operate building lighting and

HVAC equipment, and the use of automated lighting controls, which allow more efficient management of lighting energy use in commercial buildings. Manual controls that allow occupants to use equipment more efficiently can also provide significant energy savings in space heating and cooling.

5.2.4 Switching to Energy Sources/Equipment with Lower GHG Emissions

Options for switching to energy sources/equipment with lower GHG emissions include substituting solar for conventional water heaters, substituting electric for kerosene lamps, substituting biogas for other biofuel stoves, and substituting gas for coal in space heating. In many cases, the new equipment is more energy-efficient, and the new energy source has lower carbon content per unit of energy.

The use of substitute energy sources depends on their availability. In some cases, construction of energy supply infrastructure may be required (e.g., rural electrification, gas distribution network). The GHG impact of switching to electricity depends of course on the energy mix for electricity generation, and thus cannot be fully analyzed prior to conducting an electricity demand/supply integration.

In some circumstances, substituting kerosene or LPG for biofuel cookstoves may yield a reduction in GHG emissions and an increase in biomass stocks. Although kerosene and LPG are fossil fuels containing carbon, stoves using kerosene and LPG are several times more energy-efficient than typical biofuel cookstoves and have much fewer emissions of non-carbon products of incomplete combustion such as methane. A disadvantage of this option is that the petroleum products may need to be imported, while biofuels could be sustainably produced in most countries.

5.2.5 Screening Options

The applicability of different mitigation options varies among countries more in the residential and commercial sectors than in others. Options affecting space heating are not of much interest in most developing countries, while options affecting air conditioning are of low interest in the transition countries. Similarly, options affecting biomass use are not very important in the transition countries, but are of interest in many developing countries.

The general criteria for screening options are described in Chapter 2. Below we discuss several criteria that are particularly important in screening options for the residential and commercial sectors.

Size and growth rate of end use. It is important first to identify which end uses are the largest and/or the fastest growing consumers of primary energy and to target those end uses for further analysis. Surveys of equipment ownership and sales are helpful in this endeavor as are measured data on energy use of particular equipment or buildings.

New vs. existing buildings. Buildings typically have lifetimes of 50 to 100 years. In most developed nations, the annual growth rate in building stocks is relatively modest (typically 1-3%/year), which (along with the relatively long lifetime of buildings) implies that achieving significant energy savings requires retrofitting existing buildings. Such retrofits are typically more difficult and expensive than building a more efficient building in the first place. Developing nations usually have rapidly growing building stocks so programs to affect existing buildings should have a lower priority than programs and policies that target new buildings.

High cost-effective savings potential. Once the largest and fastest growing end uses are identified, one can estimate which end uses have large and cost-effective savings potentials (based on previous studies or expert judgment). Combustion-related processes are limited by the First Law of

Thermodynamics to less than 100% efficiency, while other processes that do not depend on combustion (e.g., heat pumps, cooling, and lighting) can achieve much higher efficiencies. In residential buildings, large cost-effective savings potentials are typically found in lighting, water heating, refrigeration, and space conditioning. In the commercial sector, cost-effective savings in lighting and cooling are usually large.

Ease of affecting market choices. Another important issue in screening options is the ease of changing the market to induce manufacturers to produce and consumers to purchase more efficient equipment. It is always easier to affect the efficiency of manufactured devices (such as appliances) than to affect that of site-built structures (such as most residential and commercial building shells). The quality control issues are less difficult for manufactured devices, and the number of supply-side actors to be affected by policy is usually smaller than for site-built structures. This criterion argues that for rapid short-term benefits, a focus on appliances and equipment is appropriate. It will take longer to implement standards and other policies for site-built building shells, but they are no less important in the longer term because of the long lifetime of buildings.

5.3 INPUTS FOR RESIDENTIAL AND COMMERCIAL SECTOR ANALYSIS

An analysis initially requires data on total consumption of each energy type by the residential and commercial sectors in the base year. In many countries, the division of oil, district heat, or biomass consumption between the residential and commercial sectors must be estimated. If such a disaggregation has not been done as part of the GHG Inventory, it should be estimated for mitigation analysis. In this case, analysis of options affecting these energy sources will be approximate.

A common approach for analysis of residential and commercial energy use involves disaggregating energy consumption by end uses such as space heating and cooling, cooking, lighting, refrigeration, and others. In industrial countries, analysis of technology options for specific end uses has been supported by extensive surveying by governments and utilities. The end-use approach has also been applied in developing countries for residential analysis,[2] but in many cases data are not sufficiently detailed for this approach to be used. End-use analysis of the commercial sector is particularly difficult due to the variety of building types in the sector and the general lack of survey data.

The structure of the analysis in each sector will depend on the end-use disaggregation that is chosen. For each end use considered, the minimum required inputs for scenarios are data on (1) the relevant variables that drive the demand for energy, and (2) fuel and electricity intensity.

For many end uses, the number of devices is used as a "driving" variable. For space heating, one might use floor area. For the commercial sector, where end-use analysis is more difficult, one can use total floor area (if data are available) or total valued added.

In many developing countries, it is important to distinguish between urban and rural households because the characteristics of households and buildings in these two areas can be very different. In these cases, separate analyses should be done for urban and rural households. This requires a disaggregation of energy consumption and equipment stocks in the base year into urban and rural components.

- **Energy consumption in the base year by end use.** The degree of disaggregation depends on the available data. One might choose to estimate energy consumption in the two or three most important end uses and place the remainder of energy use in an "other" category. For the commercial sector, for which end-use data are often difficult to estimate,

---

[2] Examples include India (Asian Development Bank, 1993), Mexico (Masera *et al.*, 1993), and Venezuela (Figueroa *et al.*, 1992).

one might choose to do a very simple end-use breakdown, such as dividing electricity consumption into lighting and "other," and considering all fuel use at an aggregate level.

If information on energy consumption by end use is not available from an authoritative source, then estimates must be made. Sometimes previous estimates made for countries with similar characteristics can be adopted to approximate the energy consumption by end use in the country under study. In the longer term, surveys should be done and measurements taken of who has what kind of equipment and how much energy is used by particular types of equipment. The results of these surveys and measurements can then be used to more accurately characterize the base year energy consumption by end-use.

- **Data on building/equipment stocks for each end use studied.** Data on the saturation of end-use devices (e.g., refrigerators, cook stoves) may be obtained from surveys. For space conditioning end uses, it is desirable to disaggregate the building stock into appropriate categories (such as detached houses and apartments). Similarly, for certain types of equipment, one may choose to keep separate account for sub-categories (e.g., small and large refrigerators).

- **Basic sectoral "drivers."** The basic drivers are the number of households for the residential sector and total floor area or value added for the commercial sector.[3] Projections of the future number of households often may be obtained from official sources. Projections of commercial-sector floor area may be based on growth in commercial-sector value added, which should be obtained from the macro-economic forecast used for the country study.

- **Future equipment stock for each end use studied.** For the residential sector, the usual approach is to project a future saturation of each type of equipment, which refers to the percentage of households having the equipment. The future saturation should be related to growth in household income; its estimation may be assisted by comparisons with other countries. For the commercial sector, one can use a rate of growth in the number of devices in a given end use. This rate may be based on the projected growth in floor area or value added.

- **Technology data.** The basic steps for characterizing technical options for the residential and commercial sectors are as follows (adapted from Krause *et al.*, 1987):

    - Identify the types of technologies currently in place, their average lifetimes, energy efficiency, and utilization. The degree of detail can vary considerably, from using one type/size of technology as a proxy for an entire class to developing data on many types/sizes.

    - Create an inventory of energy-efficient technologies or technical improvements and characterize their energy savings relative to current average technology. One can complement this inventory with advanced engineering options for further improving the energy efficiency of the best commercially available equipment.

---

[3] Value added is not a very desirable indicator to use because it is often not accurately measured and it is shaped by the changing structure of the commercial sector. However, one has no viable alternative if data on floor area are lacking (which is often the case).

- Assemble data on the current market prices and projected price developments for efficiency technologies or technical improvements. In considering the societal perspective, prices should be net of import duties and other taxes.

Technology characterizations for a number of important energy-efficiency technologies for the buildings sector are described in detail by Koomey *et al.* (1994).

The savings potential of a measure is typically based on engineering calculations (backed up by measured data on energy savings, in the best case) relative to a baseline technology. For space conditioning measures, these calculations typically use a building simulation computer program to estimate heating and cooling energy use in a given climate with a particular building (or an average building, known as a prototype). For other end uses, simple engineering calculations or data from equipment testing (for new equipment) usually suffice. These calculations usually compare the best currently available technology with the baseline technology.

To produce consistent economic comparisons, the analyst must make certain that the level of service delivered by efficient devices is at least equivalent to that of the devices they replace. Costs can be based on current market prices or on engineering estimates (for near-commercial technologies). Care must be used in adopting market prices because costs of efficient technology in the current market may be higher than the costs that would prevail if the market for the efficient technology grows considerably.

5.3.1 Issues in Analysis

Treating measure interactions. To properly estimate the energy savings potential of conservation measures, they should be ranked in order of increasing cost per GJ or kWh saved (the cost of conserved energy, or CCE). Determining this order is simple for efficiency measures that are independent. However, the ranking becomes complex when the energy saved by one conservation measure depends on the efficiency measures that have been implemented previously.

Consider conservation measures applied to a residential water heating system. The energy savings attributed to certain improvements in the water heater's efficiency will depend on the amount of hot water demanded, which will depend on the measures that may have already been implemented (such as low-flow showerheads). The sum of savings of each measure implemented alone will be greater than the two implemented together. If the interdependence of the measures is not taken into account, it is possible to "double-count" the energy savings.

A properly calculated efficiency potential will avoid double-counting errors by using the following procedure:

- The CCE is calculated for each of the interacting measures.
- The least expensive (i.e., lowest CCE) measure is selected and "implemented"; that is, the energy savings from the first measure are subtracted from the baseline energy use.
- The new energy use is used to recalculate the CCEs of the remaining measures. (In general, their CCEs will rise.)
- The measure with the next lowest CCE is selected and implemented.

- The energy savings of the remaining measures are recalculated and the measures are re-ranked. This procedure is repeated until all the interacting measures have been ranked.

An important consideration when analyzing commercial lighting technologies is the interaction between heating, ventilation, and air conditioning (HVAC) and lighting energy use. Reducing lighting energy will decrease cooling energy use and increase heating energy use because the waste heat from the lights will be reduced when more efficient equipment is installed. In hot climates, this effect always leads to additional energy and cost savings. The exact size of the additional savings depends on particular building characteristics and weather patterns. These interactions can be treated by simulating the effect of lighting energy savings on HVAC energy use with a building energy simulation model (Sezgen and Huang, 1994).

- **Accounting for side benefits.** Many efficient technologies provide better services than those they replace. It is important to account for these additional benefits because in some cases they can be more important than the direct value of the energy saved. An example of the importance of side benefits is that of using electric lighting as a substitute for kerosene lighting in rural homes. The gain in service (lighting quality) is generally more important than the energy savings involved. In the commercial sector, efficiency technologies may also improve worker productivity and should be allocated that benefit.

- **Avoided costs from efficiency measures.** To assess the cost-effectiveness of a mitigation option, one can compare the CCE to the avoided supply cost. One typically uses an average avoided cost. In reality, the avoided cost will vary depending on the impact of the conservation measure on the daily electricity load and the characteristics of the utility system in which the savings accrue. For example, refrigerator savings are baseload in character, occurring throughout the year at a relatively constant rate. Lighting savings are more likely to accrue at times when the utility system is faced with meeting peak demand; avoided costs will generally be higher in this case. For discussion of how to treat peak demand savings from efficiency options, see Koomey *et al.* (1990).

5.4 DEVELOPING RESIDENTIAL AND COMMERCIAL SECTOR SCENARIOS

5.4.1 A Basic End-Use Accounting Framework

This section describes a simple procedure for developing scenarios for particular end uses. For each end use selected, one projects the future evolution of equipment stocks and analyzes particular technology options for meeting future energy service demands. Since different options may be targeted at existing and new equipment or buildings, it is necessary to keep separate account of these two classes. In this method, one estimates the total number of devices (or buildings) in the target year and estimates a retirement rate for existing equipment (or buildings), which leaves a certain number surviving in the target year. The number of new devices (or buildings) that entered the stock over the study period is simply the difference between these two quantities. Efficiency measures are then applied to either existing or new buildings or equipment as appropriate.

The end uses of residential refrigeration and commercial lighting are taken as examples. In these examples, there are only two types of equipment for each end use: standard and energy-efficient. In a more sophisticated analysis, one would use more than two types.

The required inputs are an estimate of total energy consumption in the base year for the end use, a projection of equipment stocks, and average energy intensities of existing and new equipment (or buildings). In the example presented in Table 5-1, all values are expressed as indices, where the base-year (1995) value is equal to one. An analyst could use actual values rather than indices.

Step 1. Estimate the shares of base-year surviving equipment (or buildings) and new equipment (or buildings) in the future stock. To estimate a retirement rate for existing buildings or equipment, information is required on lifetimes.[4] The stock of devices (or buildings) still existing in the target year can be expressed as a fraction of the base-year stock. In the example, we assume that the total number of refrigerators grows as a function of the growth in households and in refrigerator ownership per household (saturation), while the total number of lighting fixtures in the commercial sector grows as a function of the growth in floor area and saturation (equipment per unit of floor area). The stock of new devices installed after 1995 (expressed as a fraction of the base-year stock) is simply the difference between the total stock in 2015 and the stock of devices existing in 1995 that survive to 2015.

Step 2. Use energy intensities of existing and new equipment (or buildings) to derive a baseline scenario of energy use. One can express the energy intensity of new devices as a fraction of the base-year intensities for existing devices. As illustrated in the example, the average intensity of new equipment is often lower than that of existing equipment.

In a "frozen efficiency" baseline, the intensity factors are assumed to remain constant throughout the forecast (both existing and new devices are frozen at their 1995 intensities, but new devices replace existing devices as they retire). One calculates electricity use for surviving devices by multiplying the stock of surviving equipment by the intensity of such devices and by the total energy consumption for that end use in the base year. One then multiplies the stock of new devices by the intensity of such devices and by the total electricity use in the base year, and adds the results of this and the previous calculation. This assumes that all new devices are of the standard type.

To develop a "likely trends" baseline, one must make assumptions about how intensities are likely to change because of market forces over the forecast period. The simplest approach is to assume that some share of new devices are of the efficient type (see below) rather than the average type. A more advanced approach would consider different degrees of efficiency improvement that might occur.

Equipment in buildings is usually not retrofitted (instead it is replaced at the end of its useful life) so that the assumption that surviving equipment remains at fixed intensities until retired is probably accurate. Building shells are often retrofitted due to their long lifetimes. Thus, the intensity factors for space conditioning end uses affected by shell retrofits must be adjusted to reflect such retrofits. The "frozen efficiency" baseline assumes no retrofits occur, whereas the "likely trends" baseline may assume that some of the stock receives retrofits.

Step 3. Use energy intensities of efficient technologies to derive the technical potential for energy savings. Energy use is calculated in the same manner as for the frozen efficiency case except that all new devices are assumed to be of the efficient type. If shell retrofits are being considered, one assumes that all surviving buildings receive the retrofit by the target year. The energy intensity of efficient technologies is based on the engineering calculations described earlier

---

[4] A common assumption is that devices retire at an exponential rate equal to the inverse of the device lifetime. Other retirement assumptions may be more accurate in particular circumstances.

Table 5-1: Constructing Residential/Commerical Scenarios: Two Examples

	Units	*Residential refrigerators*		*Commercial lighting*	
Lifetime of equipment	Years	20		12	
Retirement rate	%	5%		8%	
Annual growth in total energy services	%	3%		3%	
Annual growth in saturation	%		1%		1%
Annual growth in total households/floor area	%		2%		2%
Stock of devices 1995	1995 = 1.0	1.00		1.00	
Stock of devices 2015	1995 = 1.0	1.81		1.81	
Devices existing in 1995 still existing in 2015	1995 = 1.0		0.36		0.18
Post-1995 devices in 2015	1995 = 1.0		1.45		1.64
Energy intensity factor (1995 = 1.0)					
Typical existing devices 1995	1995 = 1.0	1.00		1.00	
Baseline new devices installed post-1995	1995 = 1.0	0.80		0.70	
Efficiency case new devices installed post-1995	1995 = 1.0	0.50		0.30	
Baseline scenario - Frozen efficiency					
1995 electricity use	TWh	50		40	
Total forecasted electricity use in 2015	TWh	76		53	
Electricity use from 1995 devices in 2015			18		7
Electricity use from post-1995 devices in 2015			58		46
Mitigation scenario					
Total forecasted electricity use in 2015	TWh	54		27	
Electricity use from 1995 devices in 2015			18		7
Electricity use from post-1995 devices in 2015			36		20
Annual savings in efficiency case (2015)	TWh	22		26	

(1) Devices existing in 1995 are assumed to retire at an exponential rate equal to the inverse of the equipment lifetime.

(2) Electricity use in 2015 equals intensity factor (for each vintage bin) times number of devices times 1995 electricity use.

5.4.2 Detailed Equipment Stock Models

The most accurate way to assess efficiency potentials in a bottom-up analysis is to begin at the lowest level of disaggregation that the data allow. For buildings, such analysis involves creating building prototypes with thermal and equipment characteristics corresponding to new and average existing buildings. These prototypes are then run through a building energy simulation program with average weather to estimate heating and cooling energy use for different insulation levels and equipment efficiencies. For appliances, energy use is determined from engineering simulations and measured data. This detailed approach is exemplified by the work of Krause *et al.* (1987) and Koomey *et al.* (1991). These methods require data that can be difficult to obtain so assumptions are often required.

- **Equipment stocks.** In a more detailed approach, the stock of relevant equipment is typically tracked at the level of technology, vintage (year of entry into the stock), and fuel. For example, Koomey *et al.* (1993a) describes a residential forecasting model that keeps track of stocks and other information for about a dozen different classes of refrigerators with each class allowed to choose from ten or more technologies. These stocks are inferred from industry sales data or anecdotal evidence. Building shell stocks differ by region, vintage, and space-conditioning equipment type, and in the best case these stocks are tracked at this level of detail.

  The differing lifetimes of building shells and equipment are used to estimate retirement rates and to group the building stock into different categories that are tracked explicitly. For example, buildings existing in the base year and still existing in a forecast year would be split into those with original equipment and those with equipment that has been replaced since the base year. The number of buildings in the first category would decline at a rate determined by the retirement rate of the equipment as well as the retirement rate of building shells. The number of buildings in the second category would increase over the duration of most forecasts even though the total number of base-year buildings still existing in a forecast year would decrease over time.

- **Energy intensity and efficiency factors.** Intensity factors are not usually used in the more detailed calculations. Instead, each technology is assigned an energy consumption per unit (UEC, in kWh or GJ/unit/yr), based on engineering calculations or measured data. The unit energy consumption is a function of equipment efficiency and usage behavior and these two parameters are specified independently in the most sophisticated end-use models.

  In a frozen efficiency scenario, the energy use per unit and the equipment stocks are tracked at the technology level. Total energy use in the base year is equal to the stock of equipment in that year times the average UEC for that equipment. For some devices, the capacity of the device (e.g., installed wattage for lighting) is also treated separately. The frozen efficiency forecast is created using the same conventions as before except the energy use of typical base-year existing and new devices is specified in greater detail. The forecast of equipment stocks must also be disaggregated to the technology level, which makes such forecasts data-intensive and computationally demanding.

  To estimate a business-as-usual baseline scenario, bottom-up forecasting models typically use some form of life-cycle cost calculation and diffusion curve to estimate the rate of penetration of new technology. The rate of such penetration depends on fuel prices, capital costs, logistic constraints, and other considerations.

- **Estimating technical potential.** Examples of a more detailed approach are two studies that assessed conservation potentials for residential refrigerators and commercial lighting.

  A detailed study of refrigerator efficiency options was conducted in the U.S. to determine the level at which minimum efficiency standards for refrigerators should be set (US DOE, 1989a). The analysis relies on an end-use forecasting model that characterizes refrigerator energy use and associated capital costs by class of refrigerator (e.g., automatic defrost with top-mounted freezer, manual defrost, etc.) and by technology type. The energy-savings estimates are based on engineering calculations using an established methodology. Since refrigerator usage does not vary much, these engineering calculations can give a very accurate picture of energy use for refrigerators under average conditions. The study creates a business-as-usual forecast and then assesses the potential impacts of different efficiency standard levels in terms of energy use, costs, and shipments of different technologies and classes of refrigerators.

  A detailed assessment of lighting efficiency potential is that of Atkinson *et al.* (1992). This study tracks equipment stocks at the level of technology and assesses the potential impacts of standards, incentives, and information policies on residential and commercial lighting energy use. Incandescent lamps as well as fluorescent fixtures, lamps, ballasts, and controls are treated in detail. Both current and advanced technologies are analyzed. Energy use is estimated using a bottom-up approach that relies on installed wattage, efficiency data, and usage data by technology and building type (e.g., offices, retail, restaurants, etc.). Both frozen efficiency and business-as-usual cases are presented.

5.4.3 Simple Scenario Development

This method provides a baseline scenario of energy use and a rough characterization of the potential for reducing energy use in end uses for which specific mitigation options are not analyzed. One estimates future fuel and/or electricity intensity for each of such end uses for a baseline and a mitigation scenario.[5] General knowledge of how energy is used allows a rough assessment of the likely trends. This approach relies on both consideration of historic trends and expert judgment as to how a particular end use could evolve.

The mitigation scenario represents a rough estimate of the total fuel and electricity savings that could result from a set of mitigation options. This analysis does not consider specific mitigation options. Costs of a mitigation scenario are very difficult to estimate with this method.

- **Project fuel and/or electricity intensities in each end use for a baseline scenario, and calculate total consumption of each energy type.** Projecting future energy intensities in a "likely trends" scenario requires combining historical data with judgment about the future. For example, one might assume that average energy use per household for lighting will increase from 100 kWh/year to 200 kWh/year due to increases in size of homes and in the number of lighting sockets.

  Another approach is to estimate future intensities by reference to current intensities in other countries. One makes a judgment that the end use in the study country is likely to reach the intensity level of some other country by the target year. For example, one might

---

[5] Alternately, several end uses may be grouped into a single category. For example, one might group all commercial-sector electrical end uses into a single category. Estimating changes in a category that comprises several different end uses is difficult, however.

assume that the average lighting energy intensity in the study country will reach the current level of Japan or Spain by the target year.

If many end uses are grouped in an "other" category, the change in intensity can only be estimated very roughly. For example, if one is considering total fuel use per unit floor area in the commercial sector, one might estimate that there will be a moderate increase in intensity due to higher demand for energy services such as heating or water heating.

Multiplying the intensity by the number of devices, households, or amount of floor area yields total energy consumption for each end use.

- **Project fuel and electricity intensities in each end use for a mitigation scenario, and calculate total consumption of each energy type.** The possible approaches are similar to those for projecting baseline scenario intensities except that in this case one considers changes that could occur if efficient technologies were widely adopted. If one is using actual intensities in other countries to estimate what the intensities in the study country could be, one should select a country whose residential sector or commercial sector is similar to what the same sector in the study country could be in the target year if policies encouraged efficiency improvement.

5.5 MITIGATION POLICIES

Policies affecting energy efficiency in the residential and commercial sectors fall into four basic categories: efficiency standards, incentive programs, information programs, and market transformation programs (chiefly government purchasing).

5.5.1 Minimum Efficiency Standards

Minimum efficiency standards may apply to appliances and other equipment and to new buildings or those undergoing substantial renovations. Efficiency standards can take two forms: prescriptive standards, which mandate specific technologies, and performance standards, which mandate a specific level of energy consumption. A standard can also give the user a choice of prescriptive or performance approaches.

Standards for new buildings exist in many countries throughout the world, but are often not very stringent with respect to energy efficiency (Janda and Busch, 1994). In addition, building standards are often enforced by local officials who are understaffed and insufficiently trained (Usibelli and Stevens, 1988).

A different type of standard for new buildings involves electric utilities establishing minimum efficiency requirements for new buildings hooking up to the system (Bellamy and Fey 1988). Utility service standards have been used extensively in the Northwest U.S. (Vine 1986). They have been successful in part because they apply leverage to the builder/developer at a key point in the development process. Such policies can be effective in countries where on-site enforcement may be difficult.

Because of concentration in the appliance industry and because mass-produced devices are easier to change than site-built devices, equipment standards are more effectively enforced than are building standards. Appliance standards have been popular in the U.S. (McMahon *et al.*, 1990), and are being considered in Western Europe. The data collected for the U.S. appliance standards rulemakings is extensive and has proved valuable for analysts both within and outside the U.S. (US DOE 1988, US DOE 1989a, US DOE 1989b, US DOE 1990, US DOE 1993a, US DOE 1993b, US DOE 1993c) .

By themselves, minimum efficiency standards provide no incentive to produce a more efficient device or building than the standard requires and are often much less stringent than would be cost-effective from society's perspective. They can, however, be combined with incentive programs that will encourage efficiency beyond the standard level.

Voluntary agreements between the government and appliance manufacturers can be an alternative to mandatory standards. Such agreements have been negotiated, with generally positive results, in Western Europe, Japan, and Brazil (Meyers *et al.* 1990). However, these agreements have not pushed the market as strongly as did the appliance standards in the U.S.

5.5.2 Incentive Programs

Incentives usually involve rebates or low-interest loans to those purchasing efficient devices or undertaking building retrofits. Penalties on purchasers of less efficient devices are also possible. In the U.S., rebates have been extensively used by utilities (EPRI 1987a); tax incentives have also been tried, with mixed results. In Western Europe, grants and tax incentives were used in many countries to encourage housing retrofit in the 1970s and 1980s.

Well-conceived financial incentives can influence purchaser decision-making. They may also promote the development of an energy service industry and financial infrastructure needed for continuing success with energy efficiency programs (EPRI, 1987b).

The information component of incentive programs to promote specific efficiency technologies can be important. Utility rebates for compact fluorescent lamps and electronic ballasts deliver institutional credibility to claims that these devices actually save energy. If an architect sees that the utility will pay rebates for installations of ten different devices, search costs have been reduced. The designer can just focus on ways to use those ten technologies in the design without necessarily undertaking a lengthy analysis of all techniques available to reduce energy consumption.

5.5.3 Information Programs

Information programs have been used throughout the world to promote energy conservation in homes and, to a lesser extent, in commercial buildings. Information programs include labeling of devices with efficiency information, utility bill stuffers, booklets listing the most efficient devices, handbooks for building designers, developers, architects and others, and radio and television advertisements. The impact of these programs is difficult to gauge.

Site-specific information may be delivered by energy audit programs. Without incentives or soft loans to encourage adoption of measures identified in audits, however, the rate of adoption is usually low.

The cost of disseminating information can be reduced if the information distribution source is perceived by consumers as being more objective than suppliers, is centralized (it achieves economies of scale in replication and distribution), and it uses existing distribution channels (e.g., utility bills or customer service representatives).

One important means of disseminating information is to conduct demonstration projects using the latest building and appliance efficiency technologies. These projects demonstrate the capability and reliability of new technologies. They can also uncover unanticipated problems in installing these technologies, which can then be corrected before full-scale implementation gets underway.

5.5.4 Government Purchase Programs/Market Transformation

This policy involves governments creating a market for new efficiency technology through their purchasing choices. In this way the efficient technology will be supported in the early stages, which would allow manufacturers to increase production of efficient products and achieve economies of scale more quickly. This policy can be justified on the grounds of direct operating cost-savings to the government, with the added benefit of encouraging more rapid adoption of efficiency technologies in the economy as a whole. Governments could simply mandate technologies for use in new government construction, or they could mandate a certain efficiency or performance level. Government purchase programs of this nature are being implemented at the federal level in the U.S.

5.5.5 Incentives vs. Mandates

Incentives are often preferable to mandates because they allow individual actors to choose the most cost-effective path to achieve the desired level of energy efficiency or emission reductions. Incentives are also more flexible than mandates. Mandates are useful when there are relatively high monitoring costs for incentives or when the required actions are easy, non-controversial, or inexpensive. Properly enforced mandates also may have more predictable effects than incentives although experience gained with early incentive programs can be used to great advantage when predicting the results of future programs. Mandates and incentives should work in tandem, with policy-makers using standards to eliminate the least efficient buildings or appliances and using incentives to promote improvements beyond the level of the mandated standards.

5.5.6 Effectiveness of Policies

Programs are typically only able to capture some fraction of the technical-economic savings potentials described because of real world complexities and limitations. A sophisticated attempt to address these complexities in the context of assessing savings potentials is that of Brown (1993). The results of this work, summarized in Figure 5-1, show that slightly less than half of the technical potential savings identified for the U.S. residential sector (Koomey *et al.*, 1991) can be cost-effectively captured by a combination of appliance standards, building standards, and utility programs over a twenty-year forecast time frame.

Figure 5-1 also shows the individual effect of various corrections to the estimates of energy savings and costs. These corrections each have a noticeable effect, but the biggest effect is that associated with the time necessary to scale up programs and with the imperfect effectiveness of those programs. Program costs, takeback of savings, imperfect persistence of measures, and an adjustment because engineering estimates often overestimate savings all are important correction factors that affect the technical potential. To accurately estimate the achievable potential, one should estimate the effects of all these different factors on an end-use by end-use basis. As this is a difficult process, simple estimates may be made instead.

Figure 5-1: Achievable Conservation Potential in US Residential Electricity Use in 2010

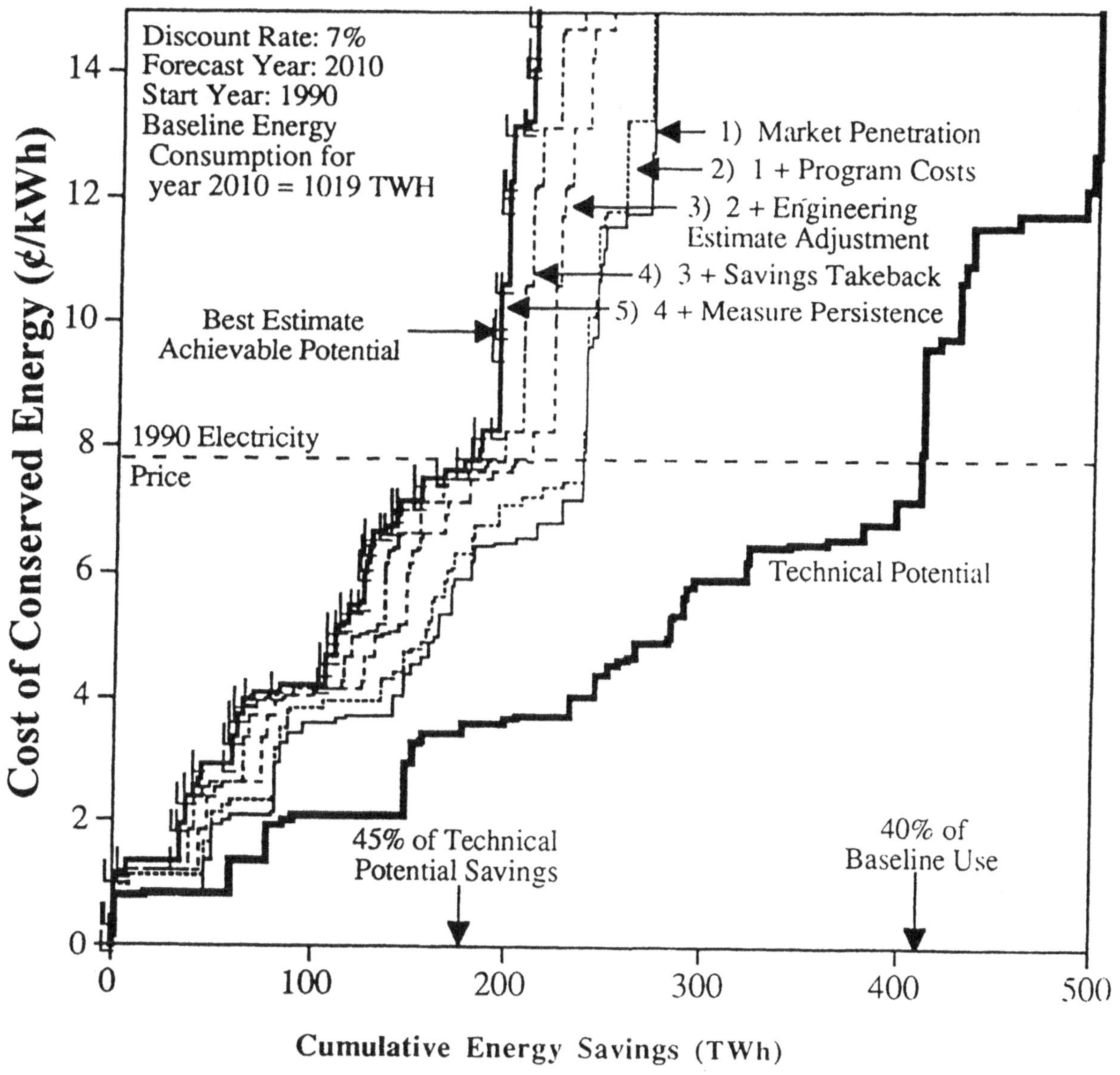

Source: Brown 1993.

REFERENCES

Asian Development Bank. 1993. Energy End Use: An Environmentally Sound Development Pathway. Manila, Philippines: Asian Development Bank.

Atkinson, Barbara, James E. McMahon, Evan Mills, Peter Chan, Terry Chan, Joseph H. Eto, Judith D. Jennings, Jonathan G. Koomey, Kenny W. Lo, Matthew Lecar, Lynn Price, Francis Rubinstein, Osman Sezgen, and Tom Wenzel. 1992. Analysis of Federal Policy Options for Improving U.S. Lighting Energy Efficiency: Commercial and Residential Buildings. Lawrence Berkeley Laboratory. LBL-31469.

Bellamy, Jim, and Jacob C. Fey. 1988. "Energy Efficient New Buildings Through a Utility Service Standard." In Proceedings of the 1988 ACEEE Summer Study on Energy Efficiency in Buildings. Asilomar, CA: American Council for an Energy Efficient Economy.

Brown, Richard E. 1993. Estimates of the Achievable Potential for Electricity Efficiency in U.S. Residences. M.S. Thesis, Energy and Resources Group, University of California, Berkeley.

EPRI, Electric Power Research Institute. 1987a. Compendium of Utility-Sponsored Energy Efficiency Rebate Programs. EPRI. EM-5579 Research Project 2884-9.

EPRI, Electric Power Research Institute. 1987b. Market Research on Demand-Side Management Programs: Preferences Among Commercial Customers. EPRI. EM-5252 Research Project 2152-2.

Figueroa, Maria, Andrea Ketoff, and Omar Masera. 1992. Residential Energy Use and Conservation in Venezuela: Results and Implications of a Household Survey in Caracas. Lawrence Berkeley Laboratory. LBL-30508.

IPCC/OECD (Intergovernmental Panel on Climate Change/Organization for Economic Cooperation and Development) 1994, "Greenhouse Gas Inventory Reference Manual IPCC Draft Guidelines for National Inventories," OECD, Paris, France.

Janda, K. and Busch, J. 1994. Worldwide Status of Energy Standards for Buildings. *Energy Policy*, Vol. 19, No. 1, p. 27-44.

Johnson, Brian J., and Catherine R. Zoi. 1992. "EPA Energy Star Computers: The Next Generation of Office Equipment." In Proceedings of the 1992 ACEEE Summer Study on Energy Efficiency in Buildings. Asilomar, CA: American Council for an Energy Efficient Economy.

Kempton, Willet, and L. Layne. 1988. "The Consumers' Energy Information Environment." In Proceedings of the 1988 ACEEE Summer Study on Energy Efficiency in Buildings. Asilomar, CA. American Council for an Energy Efficient Economy.

Koomey, Jonathan, Celina Atkinson, Alan Meier, James E. McMahon, Stan Boghosian, Barbara Atkinson, Isaac Turiel, Mark D. Levine, Bruce Nordman, and Peter Chan. 1991. The Potential for Electricity Efficiency Improvements in the U.S. Residential Sector. Lawrence Berkeley Laboratory. LBL-30477.

Koomey, Jonathan, Francis X. Johnson, James E. McMahon, Mary Orland, Mark Levine, Peter Chan, and Florentine Krause. 1993a. An Assessment of Future Energy Use and Carbon Emissions from U.S. Residences. Lawrence Berkeley Laboratory. LBL-32183.

Koomey, Jonathan, Timothy Oey, and Eric Bergman. 1993b. "The Economics of Cycling Personal Computers." *Energy Policy*. Vol. 21, No. 9. p. 937.

Koomey, Jonathan, Arthur H. Rosenfeld, and Ashok K. Gadgil. 1990. "Conservation Screening Curves to Compare Efficiency Investments to Power Plants." *Energy Policy*. Vol. 18, No. 8. p. 774.

Koomey, Jonathan G., Francis X. Johnson, Jennifer Schuman, Ellen Franconi, Steve Greenberg, Jim D. Lutz, Brent T. Griffith, Dariush Aresteh, Celina Atkinson, Kristin Heinemeier, Y. Joe Huang, Lynn Price, Greg Rosenquist, Francis M. Rubinstein, Steve Selkowitz, Haider Taha, and Isaac Turiel. 1994. Buildings Sector Demand-Side Efficiency Technology Summaries. Lawrence Berkeley Laboratory (Prepared for the Technology Characterization Database of the Intergovernmental Panel on Climate Change). LBL-33887.

Krause, Florentine, John Brown, Deborah Connell, Peter DuPont, Kathy Greely, Margaret Meal, Alan Meier, Evan Mills, and Bruce Nordman. 1987. Analysis of Michigan's Demand-Side Electric Resources in the Residential Sector. Lawrence Berkeley Laboratory. LBL-23025 (Vol. I-Executive Summary); LBL-23026 (Vol. II-Methodology and Results); LBL-23027 (Vol. III-End-use Studies).

Masera, Omar, Rafael Friedmann, and Odo de Buen, 1993. Residential Energy Use in Mexico: Structure, Evolution, Environmental Impacts, and Savings Potential. Lawrence Berkeley Laboratory. LBL-34174.

McMahon, James E., Douglas Berman, Peter Chan, Terry Chan, Jonathan Koomey, Mark D. Levine, and Steve Stoft. 1990. "Impacts of U.S. Appliance Energy Performance Standards on Consumers, Manufacturers, Electric Utilities, and the Environment." In Proceedings of the 1990 ACEEE Summer Study on Energy Efficiency in Buildings. Asilomar, CA: American Council for an Energy Efficient Economy.

Meyers, Steven, Steven Tyler, Howard Geller, Jayant Sathaye, and Lee Schipper. 1990. Energy Efficiency and Household Electric Appliances in Developing and Newly Industrialized Countries. Lawrence Berkeley Laboratory. LBL-29678.

OTA (Office of Technology Assessment) 1992. Building Energy Efficiency. U.S. Congress, Washington, D.C.

Sezgen, A.O. and Y.J. Huang. 1994. "Lighting/HVAC Interactions and Their Effects on Annual and Peak HVAC Requirements in Commercial Buildings." In Proceedings of the ACEEE 1994 Summer Study on Energy Efficiency in Buildings, Asilomar, CA:, American Council for an Energy-Efficient Economy, Vol. 3, p. 3.229.

US DOE, U.S. Department of Energy. 1988. Technical Support Document: Energy Conservation Standards for Consumer Products: Refrigerators, Furnaces, and Television Sets. U.S. Department of Energy, Assistant Secretary, Conservation and Renewable Energy, Building Equipment Division. DOE/CE-0239.

US DOE, U.S. Department of Energy. 1989a. Technical Support Document: Energy Conservation Standards for Consumer Products: Refrigerators and Furnaces. U.S. Department of Energy, Assistant Secretary, Conservation and Renewable Energy, Building Equipment Division. DOE/CE-0277.

US DOE, U.S. Department of Energy. 1989b. Technical Support Document: Energy Conservation Standards for Consumer Products: Dishwashers, Clothes Washers, and Clothes Dryers. U.S. Department of Energy, Assistant Secretary, Conservation and Renewable Energy, Building Equipment Division. DOE/CE-0267.

US DOE, U.S. Department of Energy. 1990. Technical Support Document: Energy Conservation Standards for Consumer Products: Dishwashers, Clothes Washers, and Clothes U.S. Department of Energy, Assistant Secretary, Conservation and Renewable Energy, Building Equipment Division. DOE/CE-0299P.

US DOE, U.S. Department of Energy. 1993a. Technical Support Document: Energy Conservation Standards for Consumer Products: Room Air Conditioners, Water Heaters, Direct Heating Equipment, Mobile Home Furnaces, Kitchen Ranges and Ovens, Pool Heaters, Fluorescent Lamp Ballasts, and Television Sets. U.S. Department of Energy, Assistant Secretary, Energy Efficiency and Renewable Energy, Building Equipment Division. Volume 1: Methodology; DOE/EE-0009.

US DOE, U.S. Department of Energy. 1993b. Technical Support Document: Energy Conservation Standards for Consumer Products: Room Air Conditioners, Water Heaters, Direct Heating Equipment, Mobile Home Furnaces, Kitchen Ranges and Ovens, Pool Heaters, Fluorescent Lamp Ballasts, and Television Sets. U.S. Department of Energy, Assistant Secretary, Energy Efficiency and Renewable Energy, Building Equipment Division. Volume 2: Fluorescent Lamp Ballasts, Television Sets Room Air Conditioners, and Kitchen Ranges and Ovens; DOE/EE-0009.

US DOE, U.S. Department of Energy. 1993c. Technical Support Document: Energy Conservation Standards for Consumer Products: Room Air Conditioners, Water Heaters, Direct Heating Equipment, Mobile Home Furnaces, Kitchen Ranges and Ovens, Pool Heaters, Fluorescent Lamp Ballasts, and Television Sets. U.S. Department of Energy, Assistant Secretary, Energy Efficiency and Renewable Energy, Building Equipment Division. Volume 3: Water Heaters, Pool Heaters, Direct Heating Equipment, Mobile Home Furnaces; DOE/EE-0009.

Usibelli, Anthony, and James F. Stevens. 1988. "Washington State's Energy Code for Commercial Buildings: Training and Enforcement." In Proceedings of the 1988 ACEEE Summer Study on Energy Efficiency in Buildings. Asilomar, CA: American Council for an Energy Efficient Economy.

Vine, Edward. 1986. The Residential Standards Demonstration Program: Cost Analysis. Lawrence Berkeley Laboratory. LBL-21318.

CHAPTER 6
TRANSPORTATION SECTOR

6.1 INTRODUCTION

Demand for transportation increases as economies grow, as incomes rise, and as infrastructure expands to accommodate demand. Thus, for most developing and transition countries, the key issue in analyzing transportation mitigation options is how to reduce growth in emissions from projected levels. Analysis of mitigation options needs to consider both the short term (up to a decade into the future) and the long term (a decade or longer). Unlike other energy-using activities such as industry or electric utilities, steps that can be taken to reduce emissions from transportation in the short term are either easily reversible (requiring effort to sustain the reductions), or they are of questionable value in the long term (measures to reduce energy consumption by relieving traffic congestion may be overtaken by a few years' increase in transportation demand). Other measures, such as improving the efficiency of vehicle fleets or switching fuels, require long lead times to have substantial effect, but they are much more likely to persist in their impact. Even if economic growth leads to an increase in the amount of transportation, emissions will be lower if the vehicle stock has higher fuel economy, or if it uses cleaner fuels than present petroleum fuels.

For many countries, it is important to distinguish between planning for the present transportation system and planning for one that might be expected under rapid growth. For example, if demand for transportation increases at 4% per year, and if vehicle fleets and infrastructure expand to accommodate this growth, then the transportation system at the end of a typical 30-year planning horizon will be 3.25 times larger than its present size; 69% of the system will have been built subsequent to the beginning of the planning (given replacement of existing facilities and vehicles, the percentage will be somewhat higher than this). For this reason, it is important not to let the existing system drive the analysis of mitigation options, especially in the long term, because the determinants of long-term emissions have yet to be put into place.

Transportation encompasses a complex set of activities operating at several different geographic scales even within a single country: urban, regional, national, and international. To analyze these activities, several classes of approach have been developed, each directed at different questions, different decision-makers, and different scales of analysis:

- Demand forecasting: how much travel or freight movement is expected? How might changes in fuel prices, changes in economic activity, or changes in land use affect the amount of travel or freight movement?

- Mode choice: what mix of transportation modes will be used to provide passenger and freight transportation services?

- Vehicle stock analysis: what impact does changing the vehicle technology (fuel economy, fuel type, emission controls) have on fuel use and emissions?

- Logistics management: for an activity that uses transportation, how can it be reorganized to reduce the use or cost of transportation?

- Transportation management: how should infrastructure and vehicle flow be managed to reduce congestion or to improve efficiency?

- Transportation planning: what investments should be made to meet additional demand for transportation or to improve efficiency?

All of these approaches have some value in efforts to analyze options for reducing emissions of greenhouse gases from transportation. However, most mitigation analyses will use the first three, and these will be the focus of attention in this chapter. The other approaches are more useful in analyzing options for implementation in specific contexts (e.g., a provincial agency that operates a vehicle fleet, a municipal government that builds highways or public transit systems, a development agency that builds transportation infrastructure), or for analyzing options when special circumstances apply (e.g., a government-owned fleet or a single urban area contributes a major share of the country's transportation emissions).

6.2 MITIGATION OPTIONS

Five broad categories of mitigation options encompass a large number of specific actions.

- **Improve vehicle technical efficiency**

  Reduce vehicle weight, drag, or rolling resistance; improve engine, transmission, or drive-train performance; and for some types of vehicles, switch fuels (e.g., heavy trucks from gasoline to diesel).

  Most developing and transition countries do not exercise direct control over the design of the vehicles they import or assemble (vehicles manufactured or assembled by most countries use designs if not components supplied under agreement by multinational vehicle suppliers). However, they can influence the mix of vehicles that they import, and they can negotiate with suppliers of vehicle components or vehicle designs assembled domestically to increase the fuel economy of the final product. The effectiveness of options to improve the technical efficiency of vehicles will be limited by the range of technical efficiency of vehicles that are available (or expected to be available) for import or assembly.

- **Switch to fuel systems with lower emissions**

  Shift from petroleum to natural gas, methanol, ethanol, electricity, or biofuels with lower emissions per unit of service (vehicle-kilometer traveled, tonne-kilometer of freight moved). Specific actions may include measures to ensure production and distribution of fuel, measures to ensure availability of vehicles to use the fuels, and measures to maintain the vehicles and fuel systems, with the specifics depending on the type of fuel and degree of market penetration sought. Some types of fuel switching will reduce emissions from vehicle tail pipes, but may have high enough emissions in the production and distribution stages that total emissions for the fuel cycle will be higher than for petroleum fuels. In particular, the impact of switching to electric vehicles is strongly dependent on the electricity generation resource mix.

  In general, fuel switching as a mitigation option will be limited to highway vehicles and to railroads. The latter may consider switching from coal to diesel or electric locomotives, or switching from diesel to electric.

- **Improve system efficiency**

  Improve traffic flow (e.g., improve traffic signal timing, limit mix of vehicles or access on some routes or other measures to reduce congestion or increase average speed and load); increase vehicle load factor by ride sharing, changing routes or schedules, or increased backhauling; improve vehicle maintenance; improve operating training and performance; integrate planning for expected growth. Most of these actions are subject to influence by national governments, but estimating their effectiveness can be difficult.

Mitigation options to improve system efficiency are quite diverse. Many of these options involve changing decisions made by large numbers of individuals or firms, and few have been studied extensively in terms of cause and effect. Some of these mitigation options, such as improving traffic flow, involve actions that intuitively may reduce emissions but that may do so only in the short term or not at all (Thailand Development Research Institute 1990, Birk and Zegras 1993). Only in exceptional cases, such as in Singapore, have these options succeeded (Ang 1989). Options such as training vehicle operators to improve fuel economy, or improving the maintenance of vehicles, have demonstrable effects in the short term. For example, training operators to operate their vehicles more efficiently has been shown to improve in-use fuel economy by up to 10% in the U.S. during the short run although questions remain about how long such savings will persist, or what may be required to maintain them (Greene 1986).

- **Encourage shifts toward modes with lower emissions**

  Promote walking, bicycling, public transportation, and railroad freight relative to automobile and truck traffic; avoid creating barriers to modes with lower emissions when developing, managing, or operating infrastructure for motor vehicles; increase infrastructure devoted to low-emission modes; increase the cost of using motor vehicles relative to modes with lower emissions; increase licensing requirements for motor vehicles; promote modes with lower emissions via public information campaigns or improvements in the quality of service (e.g., dedicated rights-of-way for buses, increased frequency of service). Again, estimating the effectiveness of these actions can be difficult.

- **Manage transport demand**

  Plan land uses to reduce need for motorized travel and accommodate modes with lower emissions; charge tolls for motorized traffic; improve postal and telecommunications systems; expand and decentralize public services to reduce trip length.

The theoretical attractiveness of demand management can easily be demonstrated, but the present state of knowledge provides little guidance for identifying the most cost-effective ways to reduce demand growth rates. Land-use planning and improved telecommunications have been suggested as means for reducing future levels of demand from baseline levels (Watterson 1993, Birk and Zegras 1993), but little information is available to indicate the magnitudes of reductions that might be realistically achieved.

Demand management can be affected via land-use planning to reduce the length or frequency of trips or the need to use high-emission modes for travel. Land-use planning is unlikely to have much effect in the short term, and its long-term impact is difficult to estimate. Intuitively, compact settlement patterns and mixes of land uses allow reductions in the length and frequency of trips, but numerical estimates are limited, and evidence in the U.S. suggests that "improving the mix" of land uses has relatively little impact on trip generation. Although land-use planning can provide the opportunity for people to use modes with low or no emissions, people need not avail themselves of such opportunity (Owens 1989, Giuliano and Small 1993). Long-term emission reductions resulting from changes in land-use planning may be greater in cities that have not yet adapted to extensive automobile use.

Some relationships between land-use patterns, density, and transportation remain poorly understood. Dense patterns of land use can contribute to traffic congestion, reducing average vehicle speed and fuel economy. Lower densities require longer trips and greater vehicle use, which in the long term may also lead to congestion, offsetting the short-term benefits of increased vehicle speed and fuel economy (Newman and Kenworthy, 1988).

Demand management also can be effected by substituting telecommunications or postal communications for transportation, but the impact of such measures is difficult to assess. The potential for reduction in developing countries may be small because the proportion of employees whose work and incomes would allow them to make such substitutions is low.

6.2.1 Screening Mitigation Options

In screening mitigation options, one should consider the criteria described in Chapter 2. Criteria that are especially important for transportation analysis are discussed below.

- **Relationship between an option and national goals.** Is the option consistent with development plans and goals (e.g., physical access to employment or public services; urban development)? Is it consistent with transportation development plans (e.g., planned infrastructure expansion or industrial development)? If the option is inconsistent with these plans, what are the implications and requirements of altering other plans to accommodate it? If the option is consistent with them, can some of the costs of implementing the option be shared with programs for implementing other plans? Many of the questions here relate to identifying indirect costs and benefits of the mitigation option.

- **Relationship between an option and transportation-related problems.** Does implementing the option help to solve other transportation-related problems (e.g., local air quality, traffic congestion, limited funds or land for infrastructure expansion, balance of trade for petroleum imports)? If so, does this improve its prospects for successful implementation? If not, what is required to remove the incompatibility or reduce it to acceptable levels? Many of the questions here relate to identifying direct and joint benefits of the mitigation option.

- **The country's ability to implement options.** Are there adequate numbers of technically trained personnel available to implement and manage the option? What is the cost of implementing the option and are financial resources available? Will the public support the option or oppose it? Can the needed expertise, institutions, or acceptance be developed at reasonable cost? Many of the questions here involve estimation of the direct costs of implementing the mitigation option.

6.3 INPUTS FOR TRANSPORTATION SECTOR ANALYSIS

6.3.1 Aggregate Data

- **Energy consumption in the base year.** An analysis of the transport sector initially requires data on total consumption of each energy type by each of the various transport modes (i.e., automobiles, trucks, rail, ships, air) in the base year. If information on energy consumption by mode is not available from an authoritative source, then estimates must be made.

- **Transport activity.** For passenger transport modes, the usual indicator of activity is passenger-km (p-km). The usual indicator for freight transport is tonne-km (t-km). These indicators are typically available from official sources for most, but not all modes. In some cases (e.g., private automobile travel), estimates may be necessary.

6.3.2 Vehicle Stock Data

Ideally, one has estimates of the number of vehicles in different groups defined to have similar age and fuel-type characteristics (for example, the number of vehicles that are one-year-old and use diesel, the number that are one-year-old and use gasoline, the number that are two years old and use diesel, etc.). Each of these groups is termed a "vintage." If this level of detail can be collected, then a model can be used to track a vintage as the vehicles in it are expected to age, deteriorate, or go out of service in the future. This kind of tracking is important since many of the vehicles purchased in any year will still be in use 10-15 years later. If less detail is available, larger vintages may be used with some loss in accuracy. If a vehicle fleet contains vehicles that use different types of fuels (e.g., gasoline and diesel), then if possible each vintage should be defined to include a single fuel.

Data sources will vary. Most countries impose some form of vehicle licensing, inspection, or taxation, and the administration of these requirements may collect detailed data. If enforcement and compliance with these requirements is lax, this approach will underestimate the size of the vehicle fleet, and it may be necessary to augment it with information collected for other purposes (e.g., proportion of vehicles stopped for traffic violations that are unlicensed; roadside surveys to estimate the under-count) (see Meyers 1988 for a discussion of vehicle data in developing countries). If a country imports vehicles, it may have data on past and current imports; if a country manufactures or assembles them, it may be able to obtain information on past production and disposition (sales, exports) of these vehicles. For some types of vehicles other than light-duty highway vehicles (e.g., passenger aircraft, buses, locomotives), government entities may own or control the vehicle fleet and should be in a position to obtain data directly.

6.3.3 Vehicle Fuel Economy and Use

Once the vintages have been defined and their size estimated, it is necessary to estimate how much fuel the vintage consumes now and will consume in the future. This, in turn, requires some way of estimating *vehicle use and vehicle fuel economy*. (It also requires estimating what proportion of each vintage will still be operating one year, two years, three years, and so on, in the future. This is termed the "survival rate.") For light-duty passenger vehicles, some countries may be able to obtain current vehicle use from odometer readings taken at compliance with annual licensing, inspection, or taxation; this source may also yield estimates of vehicle survival rates. Other countries may need to sample vehicle owners. If possible, it is desirable to collect additional information about variables likely to influence vehicle use, such as household income, whether the vehicle is used in a business or for government work or not, and whether the vehicle's owner is located in a large city, small city, or rural area. With this kind of information, it is possible to estimate how vehicle ownership may grow as the population increases, or as the economy expands, or as people migrate to cities.

Vehicle fuel economy estimates may be available from local assembly plants or importers although vehicle models sold in developing or transition countries often are different from those that are tested in developed countries. Sampling vehicles at refueling stations is another approach, and it may be more accurate, especially if most vehicles are operated under congested urban conditions.

6.4 DEVELOPING SCENARIOS

An approach for analyzing transportation energy demand and emissions is illustrated in Figure 6-1. The first step is to project the demand for transport for each major passenger and freight mode. There is no single, universally applicable model for projecting demand for transport. In most developing and transition countries, future transport demand is probably best projected as a function of growth in population, the economy, and income. The specific variables used will depend on the data series that are available in the country. If a country has a number of major population centers or if particular sectors of the economy contribute disproportionately to transportation demand, it may be appropriate to include information about the country's geographic or economic structure in the estimating model. The demand for particular passenger and freight modes can be estimated based on assumptions regarding the prospects for competing modes (such as rail and road in freight transport). The demand for each mode affects the number of vehicles and their rate of use.

Growth in population and the economy along with various government policies shape the levels of vehicle ownership (private and public). A vehicle stock model allows tracking of different types of vehicles as they enter or leave the stock. For each group of vehicles that enter the stock in a given year, i.e., each vintage, one collects data on basic characteristics such as fuel economy (e.g., liters per 100 km) and fuel type (e.g., gasoline, diesel). Changes in the vehicle stock over time are tracked with a vehicle stock model such as that described in Greene *et al.* (1986) or Greene and Rathi (1990).

Once a vehicle enters the stock, its fuel economy is shaped by how the vehicle is maintained and operated and also by the operating environment (i.e., the degree of traffic congestion, which affects average vehicle speeds). It may also be influenced by government policies.

Fuel consumption for each type of vehicle is calculated based on the number of vehicles, their average fuel economy, and their average use (kilometers or tonne-km). GHG and other tailpipe emissions are calculated based on fuel use and vehicle characteristics. The impact of various mitigation options is derived by estimating the effect on the appropriate variables in the model.

6.4.1 Analyzing Impacts of Mitigation Options

This section discusses how the five classes of mitigation options listed earlier might be analyzed. The analytical approaches differ, in part because the information available for analyzing them differs for each class, and in part because some draw upon types of analysis that may be undertaken for other purposes.

To analyze a mitigation option, it is necessary to estimate its effect on one or more of the key variables. For example, if an option changes the technology of vehicles that will be imported or manufactured in the future, then this will affect either the type of fuel or the fuel economy. These variables would be changed for the affected vintages and their fuel consumption recalculated. An option that changes the cost of fuel should affect the use of vehicles in all vintages, but it may do so differently for vehicles with different fuel economies or owned by people with different incomes. The level of detail and the approach used in estimating these effects will depend on the resources and expertise in each country and on how large the effect is expected to be.

If a mitigation option affects a portion of the vehicle fleet, then this needs to be analyzed by modifying the baseline estimates for the appropriate portion of each vintage. For example, if a mitigation option would reduce traffic congestion in the capital city, thereby improving fuel economy for vehicles that

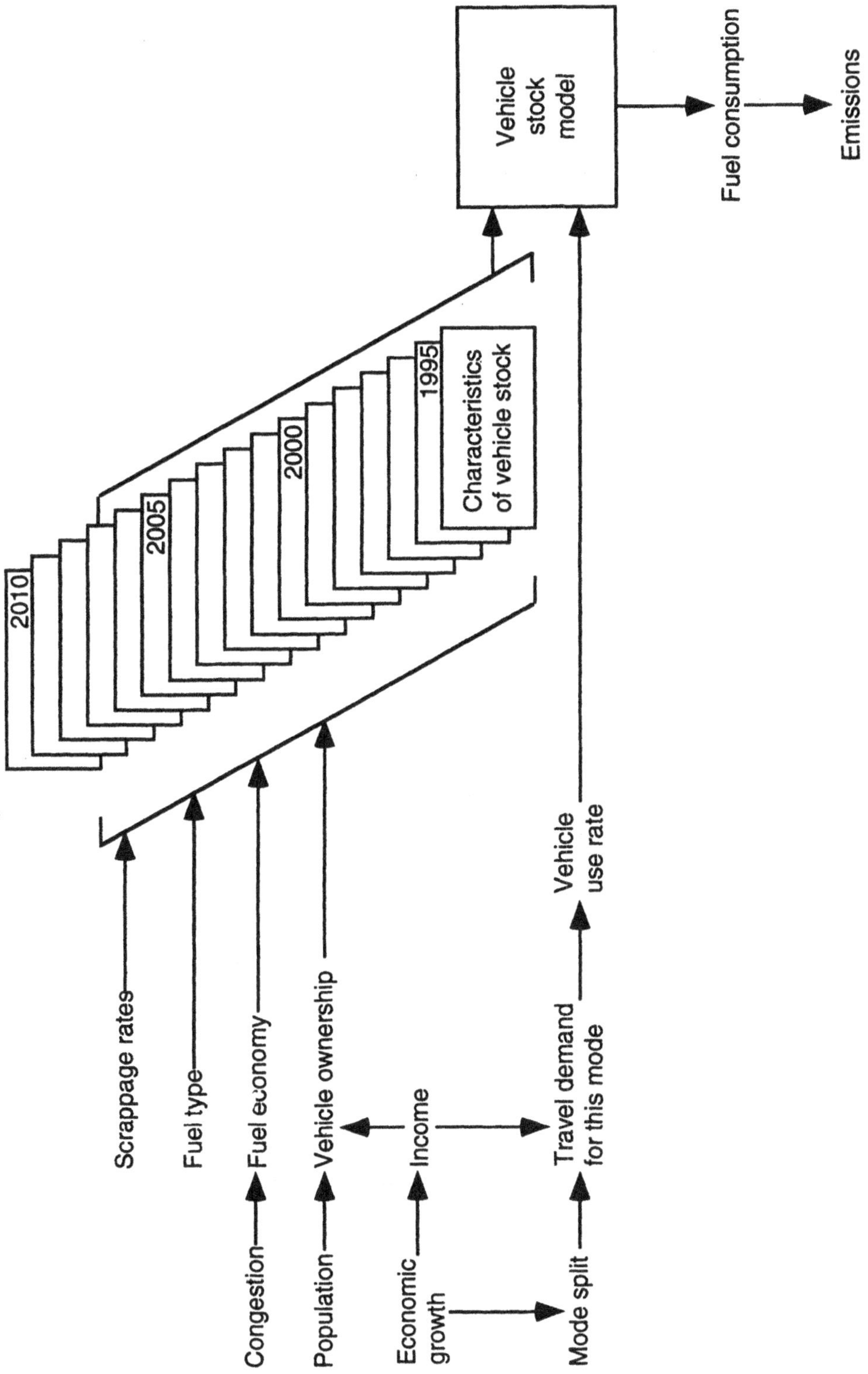

Figure 6–1: An Analytical Approach for Transportation Mitigation Assessment

operate in the capital, then the effect of the option should be estimated based on the proportion of each cohort affected by the option. Similarly, an option to encourage mode switching or reduce trip length in a major city would affect vehicle use only for portions of each vintage. An operator training program might affect only a portion of a country's operators.

Analysis of a mitigation option requires one to estimate the cost and other impacts of implementing it. Options dealing with vehicle improvements will be easier to cost than those dealing with transport system improvements.

6.4.1.1 Improve vehicle technical efficiency

The analyst must determine opportunities for improving the fuel economy of new vehicles added to the fleet (e.g., range of fuel economy available in present models available for import, assembly, or manufacture) and possible ranges of penetration rates for these vehicles (e.g., under standards the rate for efficient vehicles can be higher than otherwise). Substituting more fuel-efficient vehicles for those expected to meet future demand for vehicles in the baseline scenario allows the vehicle stock model to calculate fuel consumption and other costs and benefits over the analysis horizon. Costs will include any additional cost per vehicle associated with improved technical efficiency. Estimating the costs of different levels of fuel economy improvement would allow definition of steps on an emission reduction "supply curve."

6.4.1.2 Switch to fuel systems with lower emissions

The approach to analyzing alternative fuels requires vehicle fleet information that is essentially the same as that described above for vehicle efficiency. Typically, the penetration rates for these vehicles will be lower than those for high-efficiency vehicles using conventional fuels, at least in the early years of implementation. This reflects the added effort needed to foster the introduction of alternative-fueled vehicles in the fleet, and the perceived cost and convenience disadvantages of such vehicles relative to those that use petroleum fuels. Assessing the costs and benefits will require information about the alternative fuels (e.g., type, source, cost). This information will come from the energy supply analysis (for both conventional and renewable fuels).

A recent analysis of alternative fuels provides consistent estimates of tailpipe and upstream emissions for light- and heavy-duty highway vehicles using various alternative fuels in the U.S. (DeLuchi 1991; 1993). The upstream estimates reflect the ways the fuels would be produced in the U.S. so they would have to be modified for another country.

6.4.1.3 Improve system efficiency

To examine a mitigation option that reduces traffic congestion in order to improve fuel economy, one can estimate the effect on vehicle fuel economy, enter the revised fuel economy into the vehicle stock model, and estimate the effect of this change on fuel consumption and emissions. The effect of improving traffic flow on fuel economy (whether by adding highway capacity, improving signalization, or other means) might be estimated using traffic engineering models to estimate the effects on average vehicle speed and using coefficients to relate average vehicle speed to fuel economy. The effectiveness of the improvement, and its persistence, depend in part upon whether present congestion conditions are suppressing some demand for travel (i.e., whether or not there is latent demand that would emerge once improvements are made) and in part upon the rate of growth in demand. Both of these factors would determine how long the improvement might remain effective at reducing congestion and improving fuel economy. A program to educate vehicle operators in how to operate vehicles more efficiently would be analyzed in similar fashion.

Analysis of a program option to assist large fleet operators, including government agencies, in improving the management of their fleets would estimate the effectiveness of such programs in reducing distance driven or improving fuel economy, apply these effects to the proportion of the nation's vehicles affected, and calculate the fuel savings. Experience from the U.S. and Canada suggests that improved routing and logistics management for large fleets can reduce fuel consumption by 10-20% (Erkut and MacLean 1992, Graham 1993). Persistence of savings from this option will depend in part upon whether the option is implemented as a one-time program or on a continuing basis, and in part on economic incentives to maintain efficient performance (either via fuel prices or competition with other firms).

For both operator training and fleet management, the effect on a country's emissions would depend on the proportion of the nation's operators and vehicles affected by the options, as well as on the effectiveness of the options among the targeted populations.

This category also includes options to increase vehicle occupancy (e.g., by increasing the opportunities available for freight backhauling). These options would be modeled as reductions in the rate of vehicle use for the affected vehicles.

6.4.1.4 Encourage shifts toward modes with lower emissions

Mode shifting of passenger travel, especially in urban areas, entails encouraging people to use non-motorized modes or public transportation instead of automobiles. Analysis of this option requires estimates of present and projected modal splits and estimates of emissions intensities for various modes. The analysis must consider any changes in the operation of low-emission modes taken to respond to changes in demand (e.g., an increase in the frequency of bus service or an increase in the area served by buses). Absent an aggressive effort to improve the performance of low-emission modes and to encourage their use, the impact of such mitigation options is likely to be fairly small. Again, the technical assistance team may provide some estimates of the effectiveness of mode-switching measures, based on international experience.

The analysis task may be easier for freight when an established alternative to shipping by truck (e.g., by rail) exists.

6.4.2 Constructing Scenarios: An Example

6.4.2.1 Baseline scenario

Fuel use in a baseline scenario may be projected simply by assuming continuation of present rates of increase and augmenting this information with expected changes (e.g., planned infrastructure investments, anticipated effects of policies recently enacted but not yet fully implemented). Alternatively, expected fuel use may be projected in more detail based on expected trends in vehicle ownership and technology (e.g., fuel economy for vehicles entering the country's fleet, absent any measure to improve it). This is the method used in the baseline scenario presented in Box 6-1.

To calculate how much fuel each vintage of vehicles will consume in a given year, one multiplies fuel economy (liters per kilometer) times the number of vehicles times the average number of kilometers a vehicle is expected to be driven in a year. Summing fuel consumption across vintages yields total fuel consumption.

To construct the baseline scenario, it is necessary to develop information on vintage size, fuel economy, vehicle use, and survival rates for the vehicles that will enter service in future years. The simplest approach is to extrapolate from past vehicle sales and to assume similar fuel economy and patterns of use. However, vehicle purchase rates are known to vary with the health of the economy, with growth of personal income, with the availability of vehicles if imports or production have been restricted in the past, and with changes in government policy toward vehicles. These same factors are also known to influence how much vehicles are used. If there are reasons to believe that future rates of vehicle purchase or use will differ substantially from those in the past, then these rates need to be modeled.

The fuel consumption for each vintage of new vehicles can be calculated as it is for vintages of existing vehicles, and these new vintages can be tracked from year to year. Their estimated fuel consumption in each year is added to that estimated for existing vehicles.

The basic approach outlined above can be applied to different types of vehicles. For freight modes, tonne-kilometers may be a more appropriate measure for vehicle use than distance. For rail, ship, and air transport, energy use can be projected without a vehicle stock analysis. One can estimate growth in demand for passenger and freight service based on expected trends and then estimate change in the average energy intensity (e.g., energy use per passenger-km or tonne-km).

Box 6-1. Constructing Transportation Scenarios Using a Vehicle Stock Model (Modified from Greene *et al.*, 1986). Base year = 1995.

Distribution of vehicles by vintage in Country X (light trucks):

Year in Service	Number	Year in Service	Number
1995	8,482	1986	1,036
1994	8,064	1985	967
1993	5,871	1984	1,209
1992	5,180	1983	691
1991	4,489	1982	760
1990	4,248	1981	587
1989	3,212	1980	380
1988	2,625	1979 or earlier	1,140
1987	1,450	TOTAL	50,391

Use of new vehicle in first year of operation: 54,501km/yr (estimated from survey).
Rate of decline in use with age: -0.036 (estimated from survey).
Average fuel economy for existing vehicles: 9.24 litres/100km (99% of fuel used is diesel).
Scrappage function estimated from survey.

Baseline scenario assumptions (projection through 2003):

Average new vehicle fuel economy: 9.5 litres/100km
(estimated from survey; assume constant for new vehicles).

Annual growth rate in vehicle stock: 14%/yr (estimated from vehicle registration data)

Registrations

	New vehicles	Used vehicles
1996	9,669	819
1997	11,023	933
1998	12,566	1,064
1999	14,325	1,212
2000	16,331	1,383
2001	18,617	1,576
2002	21,224	1,797
2003	24,196	2,048

No change in real prices of vehicles or fuel through 2003.

Box 6.1. (Continued)

Mitigation scenario assumptions:

Average vehicle fuel economy: improve tires, maintenance, and engine oils; institute driver training to yield an immediate improvement of 5% for existing and new vehicles; otherwise fuel economy of new vehicles does not change.

Impose 11% tax on new vehicles: reduces purchases of new vehicles by 10% in 1997; growth of registrations declines to 10% per year.

Registrations

	New vehicles	Used vehicles
1996	9,669	819
1997	10,635	933
1998	11,699	1,064
1999	14,325	1,212
2000	12,869	1,383
2001	14,156	1,576
2002	15,572	1,797
2003	17,129	2,048

Increase diesel fuel price by 10% in 1996 and by 30% in 1997; increase price of gasoline by 12% in year 1996 and 14% in year 1997 to reduce incentive to purchase new gasoline vehicles instead of diesel. The price increase reduces driving via an elasticity parameter.

Results:

	1995	2003 Base Case	2003 Mitigation
Vehicle stock	50,391	124,000	97,000
Fuel consumption (diesel) (10^6litres/yr)	183	565	379

6.4.2.2 Mitigation scenario

Estimating the impact of mitigation options on energy consumption and emissions requires assumptions regarding the rate of implementation. As noted above, this estimation will be difficult for some classes of options.

One may evaluate specific actions either in isolation from or in combination with each other. For example, the mitigation scenario presented in Box 6-1 combines the effects of several measures to reduce the rate of growth in vehicle purchases, plus measures to improve the efficiency of vehicle operations. Estimates of cost are likely to be more difficult for combined actions than for individual ones. The uncertain costs and effectiveness of some classes of options, such as demand management, may make their use in supply curve development of questionable value.

Estimation of macro-economic impacts of mitigation options (e.g., on employment or productivity) would require additional information on the manner and cost of implementation, plus the benefits of implementation. For example, in the mitigation scenario in Box 6-1, future vehicle registrations are substantially reduced when compared to the baseline scenario. If these vehicles are used primarily for convenience, the macro-economic effects of the scenario would be quite different than if the vehicles were used productively in the economy.

6.5 MITIGATION POLICIES

- **Improve vehicle technical efficiency**

  A number of policy options are available for improving the technical efficiency of vehicles. Where vehicles are imported, the most effective policy is likely to be *fuel economy standards*. In many countries, import duties and other taxes make vehicle purchase costs so high that adding fees based on fuel economy are unlikely to have much effect. Where vehicles are assembled or manufactured domestically using designs or components from abroad, negotiations probably will be necessary to gain access to models with higher fuel economy. These might be augmented with a gradually increasing schedule of fuel economy requirements. As noted earlier, efforts to improve vehicle fuel efficiency are likely to be limited to whatever range the multinational vehicle manufacturers have developed for their global markets.

- **Switch to fuel systems with lower emissions**

  Fuel switching is a more complex undertaking because it requires coordination among the different activities of vehicle production, vehicle purchase, fuel production and distribution, fuel purchase, and vehicle maintenance. It typically requires a mix of measures to ensure: production of vehicles that can use the alternative fuel; provision of these vehicles at a price competitive with those that use conventional fuel; production and distribution of the alternative fuel where it is demanded; provision of the fuel at a price competitive with conventional petroleum fuels; conversion of existing vehicles to use the new fuel; and maintenance and repair of vehicles that use the new fuel.

  A fuel-switching option is likely to require a mix of: *requirements* for fleets to purchase the vehicles; *mandates* for production or import of these vehicles in some share of the market; *subsidies* during the initial phase of the program for vehicle conversion and fuel production; some sort of *tax or rebate* to provide price parity among fuels; and *training and licensing programs* for mechanics to convert or maintain vehicles that use the new fuel. The effectiveness of these measures will depend on the mix and the aggressiveness with which they are pursued.

- **Improve system efficiency**

  Some options to improve traffic flow involve changes to transportation infrastructure, which historically has been a responsibility of the public sector. These options can be implemented relatively easily by *direct investment of public funds*, assuming the availability of funds for changing and maintaining traffic signalization systems, expanding highway capacity, providing dedicated rights of way for buses or car pools, and so forth. More controversial and more difficult to implement are *bans* on certain types of traffic (such as trucks) during periods of peak congestion, or *restrictions* on the numbers of vehicles or vehicle licenses. Some of these measures, if implemented, require supporting measures to provide or improve alternative modes.

- **Encourage shifts toward modes with lower emissions**

  The choice of measures to encourage the use of modes with lower emissions will depend upon the present availability and use of these modes, and on understanding present barriers to their increased use. For example, if alternatives are presently available but under-used, surveys can identify reasons for the lack of use, each of which would require different measures (for example, *investment* to improve service, *changes in prices* of different modes) to correct. On the other hand, if alternatives are used to present capacity, then *investment in new capacity* may be the most effective measure to encourage greater use. Other measures, depending upon specific barriers identified, might include a *shift in funding emphasis* toward modes with lower emissions, *higher fees* for modes with higher emissions, and *managing land-use change* to promote patterns of use and infrastructure that are more easily and effectively served by modes with low emissions.

- **Manage transport demand**

  Measures to manage transport demand will vary with local circumstances. In some countries, lack of reliable telephone and postal service requires clients and couriers to make large numbers of trips that businesses and households in other countries do not have to make. *Investment in postal, telecommunication, and service delivery systems* targeted to foster the use of these systems instead of physically moving information and people to conduct business or household functions could reduce travel and, with postal systems, provide employment. Implementation of *land-use planning* that could reduce travel demand requires coordination among the responsible agencies and a solid understanding of the reasons for present travel behavior.

REFERENCES

Ang, B. 1989. "Traffic Management Systems and Energy Savings: The Case of Singapore," Department of Industrial Systems Engineering, National University of Singapore, Singapore.

Birk, Mia Layne and Christopher Zegras. 1993. Moving Toward Integrated Transport Planning: Energy, Environment, and Mobility in Four Asian Cities. Washington: International Institute for Energy Conservation.

DeLuchi, M. A. 1991. Emissions of Greenhouse Gases from the Use of Transportation Fuels and Electricity. Report ANL/ESD/TM-22, Vol.1. Argonne, IL: Argonne National Laboratory.

DeLuchi, M. A. 1993. Emissions of Greenhouse Gases from the Use of Transportation Fuels and Electricity. Report ANL/ESD/TM-22, Vol.2. Argonne, IL: Argonne National Laboratory.

Erkut, Erhan and Doug MacLean. 1992. "Alberta's Energy Efficiency Branch Conducts Energy Audits," *Interfaces*, Vol. 22, No 3, May- June, pp. 15-21.

Giuliano, Genevieve and Kenneth A. Small. 1993. "Is the Journey to Work Explained by Urban Structure?" *Urban Studies*, Vol. 30, No. 9, November, pp. 1485-1500.

Graham, Derek S. 1993. "A GIS for Bus Routing Saves Money, Worry in North Carolina," *Geo Info Systems*, Vol. 3, No. 5, May, pp. 39- 43.

Greene, D. L. 1986. Driver Energy Conservation Awareness Training: Review and Recommendations for a National Program. Report ORNL/TM-9897. Oak Ridge, TN: Oak Ridge National Laboratory.

Greene, David L., Nabil Meddeb and Jin-Tan Liu. 1986. "Vehicle stock modeling of highway energy use: Tunisian and US applications," *Energy Policy*, Vol. 14., No. 5, October, 1986, pp. 437-446.

Greene, David L. and Anju Rathi. 1990. Alternative Motor Fuel Use Model: Model Theory and Design and User's Guide. Report ORNL/TM-11448. Oak Ridge, TN: Oak Ridge National Laboratory.

Meyers, S. 1988. Transportation in the LDCs: A Major Area of Growth in World Oil Demand. Report No. LBL-24198, Berkeley, CA: Lawrence Berkeley Laboratory.

Newman, Peter W. G. and Jeffrey R. Kenworthy. 1988. "The Transport Energy Trade-Off: Fuel-Efficient Traffic Versus Fuel-Efficient Cities," *Transportation Research A*, Vol. 22A, No. 3, May, pp. 163-174.

Owens, Susan. 1989. "Models and urban energy policy: A review and critique," pp. 227-244 in L. Lundqvist, L-G. Mattesson and E.A. Ericsson (eds), Spatial Energy Analysis, Aldershot, UK: Avebury.

Thailand Development Research Institute 1990. Energy and Environment: Choosing the Right Mix. Jomtien, Chonburi: Research Report No. 7, Presented at the 1990 TDRI Year End Conference, "Industrializing Thailand and Its Impact on the Environment".

Watterson, W.T. 1993. "Linked simulation of land use and transportation systems: Developments and experience in the Puget Sound region," *Transportation Research A*, Vol. 27A, No. 3, May, pp. 193-206.

CHAPTER 7
AGRICULTURAL SECTOR - ENERGY USES

7.1 INTRODUCTION

The agriculture sector has at its core the production process for foodstuff (e.g., grains, fruits and vegetables, meat, fish, poultry, and milk), and non-food vegetable products of economic value (e.g., tobacco, jute, hemp). However, the sector also comprises or has close links with processes that take place before and after this core production process, such as fertilizer production, post-harvest processing, and transport of foodstuff. Defined broadly, the agriculture sector has as its primary goal the delivery of food on the table for the population or for export. Thus, any measure that will reduce the energy consumption while delivering the food service is in principle a potential candidate for analysis as a mitigation option.

Despite the relative importance of this sector to economic activity and employment in the developing countries, agricultural energy use tends to be small compared to that in industry or transport. Energy is mainly used for ground water pumping and farm machinery such as threshers and tractors. In many cases, electricity and fuel use tends to be inefficient because of price subsidies, and thus mitigation options may offer a significant potential for improving efficiency and reducing GHG emissions from this sector.

7.2 MITIGATION OPTIONS

Potential mitigation options for agricultural energy use are described below. While some of the options are not yet available for widespread implementation, or need more scientific and economic analysis before their applicability can be assessed, they are also presented since they might become feasible later on.

The main near-term option likely to be of interest for GHG mitigation is efficiency improvement in irrigation. The use of various renewable sources of energy for agricultural applications (e.g., wind-driven pumps, solar drying, diesel engines powered with mostly gasified biomass) have been tested on a limited scale and may be of interest in some cases. (Agricultural residues may also be used for meeting energy demands outside the agriculture sector - e.g., for cogeneration in agro-processing industries.)

- **Reduce energy use for irrigation.** Irrigating crops often requires considerable amounts of electricity or diesel fuel. Reducing energy consumption for irrigation while providing the desired service may be accomplished through use of more efficient pump sets and water-frugal farming methods.

  To improve the efficiency of irrigation pump sets, a number of technical measures are available. These include: use foot valves that have low-flow resistance; replace undersized pipes and reduce number of elbows and other fittings that cause frictional losses; use high-efficiency pumps; select pumps better matched to the required lift characteristics; use rigid PVC pipes for suction and delivery; operate pumps at the recommended RPM; select prime mover for the pump (i.e., electric motor or diesel engine) matched to the load; select an efficient diesel engine or motor for the application; schedule and perform recommended maintenance of the pump and the prime mover; and ensure efficient transmission of mechanical power from the prime mover to the pump. For further discussion of the potential for improving pump sets, see OR Group (1993).

- **Increase the efficiency of non-pumping farm machinery.** Energy use for traction for cultivation, sowing, weeding, harvesting, and other operations can be reduced through use of more efficient equipment or by minimizing the need for traction through low-tillage agriculture.

- **Switch to lower-carbon energy sources.** Options in this category include wind- and photovoltaic-powered pumps, enhanced solar drying, and use of biofuels instead of fossil fuels in various applications where heat is required.

- **Reduce input of chemical fertilizers.** The two basic ways of reducing the input of chemical fertilizers are to target fertilizer application better and to substitute organic or microbial fertilizers for chemical fertilizers. Reduced demand for chemical fertilizer lowers energy use in the chemical industry. There have been limited studies in developing and transition countries on reducing the intensity of chemical fertilizer inputs through improved application or use of organic fertilizers so assessing the potential impact of this option is difficult.

- **Use conservation tillage systems.** Conservation tillage practices store carbon in the soil through retention of vegetative matter (crop residue). Since most conservation tillage practices reduce the number of trips across a field needed to grow and gather a crop, total energy required to grow a crop is reduced.

- **Improve efficiency of post-harvest drying and storage.** Various agricultural products are subjected to drying or cold storage before they are sent to market. The efficiency of these processes can generally be improved through use of better equipment and proper maintenance.

- **Reduce post-harvest foodgrain losses.** Assuming that food needs are being met, use of storage methods impervious to pests and rodents can reduce the need for crop production, thereby saving the energy that would be used in that production.

7.2.1 Screening of Options

Screening of options for the agriculture sector should consider their possible impacts on a number of critical issues other than cost-effectiveness and reduction in GHG emissions. Analysis of these impacts is important because some of them may be unacceptable despite attractive GHG reduction potential for the option. Some of the most significant issues for screening (and evaluating) mitigation options are:

- Impact on the local environment
- Impact on unemployment and on enterprise development
- Impact on income distribution and social equity
- Impact on national security - dependence on imports for food, fuel
- Impact on capital investment
- Impact on foreign exchange reserves (i.e., balance of payment).

Each country may have specific additional issues that require consideration in the screening of mitigation options. Some of these issues have quantitative criteria for acceptability (e.g., the rate of return on social investment must meet the norm set for other social investments). Others may have only qualitative criteria.

7.3 INPUTS FOR AGRICULTURE SECTOR ANALYSIS

The initial inputs for agricultural scenarios are base-year data on agricultural production and fuel and electricity use. A projection of future value added or the future production of food grain can be based on government plans or expert estimates. The analyst may divide the agricultural sector into two components: irrigated agriculture, where data and information on particular mitigation options for pumping and on-farm machinery may be available, and non-irrigated area, whose commercial energy use and intensity is usually very low.

Analysis of selected equipment requires disaggregation of energy consumption by end use. Major end uses include irrigation, field operations, post-harvest processing, and livestock facilities. Detailed end-use data are often sparse for the agriculture sector. As a rough approximation, one can usually assume that nearly all of the electricity consumption is for irrigation. Fuel use is divided primarily between diesel pumps and field operations.

End uses being analyzed should be characterized in terms of the number and average energy intensity of existing equipment. Projection of the future stock of each type of equipment may be based on future production. For projecting the stock of pumping equipment, estimates of the future amount of irrigated crop area and future crop patterns are also useful inputs.

An analysis of new equipment requires similar data as in other energy-demand sectors. For example, data on performance and cost of standard and energy-efficient pump sets may be gathered from equipment manufacturers, and the cost and energy savings may be calculated for one or more typical applications. One may use one size/type of equipment or several.

For retrofit improvements to existing equipment, the average energy savings and cost for typical applications may be based on actual measurements from projects or on engineering calculations.

7.4 DEVELOPING SCENARIOS

7.4.1. An End-Use Accounting Framework

The basic steps in an end-use analysis of agricultural equipment are listed in Table 7-1. This analysis requires an estimate of base-year electricity or fuel consumption for the end use and considers evolution of equipment stocks. For irrigation, for example, one begins with base-year data on the number and average energy intensity of pump sets by type (fuel/electric) and size and then projects growth in the total number of pumps of each type/size. One retires existing pumps at a specified rate and these are replaced by new pumps. Mitigation options that may be considered include use of high-efficiency rather than standard pumps and retrofit of existing pumping systems.

Where agricultural energy consumption is a significant component of total energy use, a study team may wish to use a more detailed approach which analyzes the evolution and energy intensity of irrigation pumps and other equipment for each crop type (see Moulik *et al.* 1990). This approach requires sufficient data and information on irrigation pumps and other equipment to be able to project their numbers and future energy intensity with some confidence.

The cost-effectiveness of technology options can be based on how the CCE compares to the cost of electricity or fuel. If a measure meets the test of economic acceptability from the societal perspective, one should assess its impact on other important issues (such as unemployment and enterprise development, local environment, food security, foreign exchange reserves).

Table 7-1. Basic End-Use Analysis for Agriculture Energy (Electric or Diesel Irrigation Pumps Example)

- Obtain data on base-year number and average energy intensity of pumps by size.
- Estimate future number of pumps in each size class.
  – Use growth rate based on projected growth in VA or irrigated area.
- Estimate retirement rate for base-year pumps.
- Obtain data on average energy intensity and cost of new standard and high-efficiency pumps.
- Estimate potential average energy savings and cost associated with retrofit of surviving base-year pumps.
- Calculate *baseline* energy consumption for each class of pumps.
  – Frozen efficiency: no retrofits, standard new pumps
  – Likely trends: some retrofits, mix of standard and high-efficiency new pumps
- Calculate *technical potential* energy consumption.
  – ~100% retrofits, ~100% high-efficiency new pumps*
- Estimate portion of technical potential that is cost-effective.
- Identify and evaluate mitigation policies and estimate achievable potential.

* In practice, the measures will not be applicable for all cases; some adjustment must be made.

7.4.2 Scenario Development Without End-Use Analysis

If data are lacking to conduct end-use analysis, a simple approach may be used to construct scenarios for agricultural energy use. The steps in an aggregate analysis are listed in Table 7-2. This analysis uses a measure of total agricultural production such as the total tonnes of food grain or total value added. The main task is to project the future aggregate fuel and electricity intensity (i.e., energy use per tonne of grain or unit of value added) for baseline and mitigation scenarios. The technical potential for GHG mitigation is roughly assessed without evaluation of specific options.

Table 7-2. Aggregate Analysis for Agriculture Energy Use

• Obtain base-year grain production or value added by agriculture.
• Obtain or estimate base-year energy use for agriculture by fuel type.
• Project future grain production or value added by agriculture using likely trends or government plans.
• Calculate the base-year intensity of fuel and electricity use.
• Project future baseline intensity of fuel and electricity use.
• Project future intensity of fuel and electricity use in mitigation scenario.
• Calculate future fuel and electricity use.
• Estimate mitigation scenario fuel and electricity use.
• Estimate fuel and electricity savings in mitigation scenario.
• Identify mitigation policies.

The construction of baseline and mitigation energy scenarios at the aggregate level is relatively straightforward. The baseline projection of energy intensities should consider past trends and also what areas of land will be under cultivation for various crops, using what degree of irrigation, machinery, and fertilizers. The projection of aggregate fuel and electricity intensity in the mitigation scenario should be done by roughly estimating the technical potential for reducing intensities. For example, one might judge that a variety of options could reduce aggregate electricity intensity by 15% in the target year. In this approach, estimating the cost of mitigation is very difficult.

7.5 MITIGATION POLICIES

The available policy options for the agriculture sector are generally similar to those for industry. They include efficiency standards for generic equipment, energy audits, loan programs, and financial incentives. In both sectors, opportunities for efficient use of resources and use of non-fossil energy resources tend to be rather site-specific so policy responses require some kind of outreach. Agriculture presents a particular challenge, however, because of the large number of individual enterprises spread over a wide area. Thus, greater management resources are needed to implement and monitor policies. In addition, in many countries, subsidies to energy and other agricultural inputs limit interest in their more efficient use. Reduction or removal of such subsidies will encourage both more efficient use and application of renewable technologies.

Design of policies needs to account for the impacts on different stake-holders, especially those who may stand to lose from a policy and will thus need to be rewarded (Bhatia and Pereira, 1988). For example, reduced electricity sales from more efficient electric pump sets might be attractive to a utility because of reduced outlays for subsidy of agricultural electricity. On the other hand, the farmer may have higher payments of principal and interest. Similarly, there will be economic impacts on technology suppliers, consumers, traders, and so on. If the measure is societally attractive but not attractive from the perspective of some of the stake-holders, it need not be rejected. Policy means, such as easy credit, may be used to compensate the losers while the society as a whole still gains from the implementation of options.

REFERENCES

Bhatia, R. and Pereira, Eds. 1988. Socio-economic Aspects of Renewable Energy Technologies. Praeger, New York.

Moulik, T., Dholakia, B. and Shukla, P. 1990. Energy Demand Forecast for Agriculture in India. *Economic and Political Weekly*, Dec. 29.

OR Group. 1993. Surveys of Pump Set Rectification Potential and Electricity and Diesel Use. Baroda, India.

CHAPTER 8
CONVENTIONAL ENERGY SUPPLY

8.1 INTRODUCTION

This chapter discusses the identification and screening of mitigation options in conventional energy supplies, resource assessment and technology characterization, and the policies that might be used to implement options considered desirable. Comparative analysis of mitigation options affecting the conventional energy supply system and development of energy supply scenarios may be conducted with one of the bottom-up models discussed in Chapter 3, and not described in this chapter.

The conventional energy supply system consists of the following components:

- Oil sector
- Gas sector
- Coal sector
- Nuclear materials sector
- Electric power sector

Although the electric power sector is often the largest contributor to GHG emissions, all elements of the fuel cycle need to be considered for their mitigation potential. Table 8-1 identifies some of the GHG emission sources from the conventional energy supply system.

The application of GHG mitigation options to the conventional energy supply system is a major undertaking. The options affect large energy supply facilities (e.g., power stations, coal mines, gas distribution systems); they are complex to apply because of the highly interconnected nature of the energy supply system; and they can have major impacts on the overall economy.

The process of analyzing GHG mitigation options for the conventional energy supply system builds on the well-developed techniques of energy system planning. The addition of environmental analyses to these energy planning techniques is relatively recent but has received much attention. The addition of GHG emission calculations and the evaluation of mitigation options is an incremental analysis that can added to normal energy system planning methods.

8.2 MITIGATION TECHNOLOGY OPTIONS

GHG analyses are typically done over a long time horizon. Over a long time period, conventional energy supply systems are likely to change dramatically and so will potential GHG mitigation options. In addition to mitigation options that are commercially available at this time, this section identifies and characterizes technologies that are at various different development stages and may, at a later time, become available.

Table 8-1. Sources of GHG Emissions from the Conventional Energy Supply System

Sector/Fuel Cycle Stage	Source of Emissions	Greenhouse Gas Emitted CO_2	CH_4	CO	NO_x
OIL SECTOR					
Production	Gas flaring	X	X		
Transport	Spills		X		
Refining	Distillation Fractionation Spills Storage Leaks Combustion	X	X	X	X
GAS SECTOR					
Production	Gas flaring Leaks		X		
Transport	Pipeline leaks		X		
Liquefaction/ regasification	Leaks		X		
COAL SECTOR					
Mining	Coal bed methane		X		
Transport					
Cleaning		X	X	X	
NUCLEAR MATERIALS SECTOR					
Mining		X			
Processing		X	X	X	X
ELECTRIC POWER SECTOR					
Generation	Combustion	X	X	X	X

For the power sector, mitigation options are listed in three areas: technology and efficiency improvements, fuel substitution, and post-combustion techniques. Mitigation options applicable to the entire energy supply system, including resource extraction and transportation/distribution, are also presented. Only a brief summary of each option is presented here. More details, including data on GHG emission reduction potential and cost, are described in the *IPCC Technology Characterization Inventory* (IPCC, 1993).

Not all of these options will be applicable in every country. In addition, countries may wish to evaluate other mitigation options not presented here. It will be necessary to screen the available options and select the most appropriate ones for detailed analysis.

8.2.1 Power Sector: Technological and Efficiency Improvements

- **Coal Beneficiation**

  The purpose of coal beneficiation (coal preparation or coal cleaning) in the energy sector is to remove impurities such as ash-forming minerals and sulfur to improve the combustion characteristics of the coal. There are three basic technology types of coal beneficiation: physical beneficiation (commercially available), chemical cleaning (under development), and biological cleaning (under development).

  Physical coal cleaning employs technologies based on physical differences between the coal and the mineral impurities. During the process, the coal is crushed and screened into different size categories to separate out impurities that are not chemically bound to the organic matter of the coal. The coal fines may either be discarded or cleaned using froth flotation. In the U.S. alone, there are several hundred beneficiation facilities in operation using physical coal cleaning. Chemical cleaning uses chemical reactions to remove impurities (e.g., sulfur) that are organically bound to the carbon in the coal. Biological beneficiation employs microbes specifically designed to selectively attack and break down impurities in the coal. Both chemical and biological cleaning have the potential for significantly higher removal efficiencies.

  Benefits of cleaning the coal prior to combustion include (1) improved boiler availability and reduced maintenance, (2) reduced SO_2 and dust emissions as well as significantly lower waste generation, and (3) increased heat content of the product coal. A higher heat content of the coal corresponds to lower CO_2 emissions per kWh electricity generated.

  Costs of coal beneficiation depend on the coal feedstock and the level of cleaning. Capital costs for physical cleaning can range from $25,000 to $100,000 per tonne/hr. Related O&M costs range from $1-5 per tonne of cleaned coal (IPCC, 1993).

- **Advanced pulverized coal combustion**

  Significant research and development efforts are directed toward improving operating characteristics of conventional pulverized coal-fired (PC) power plants. Plant efficiencies of close to 40% are targeted that will be achieved through the use of supercritical steam, higher initial steam temperatures, and multiple reheat. Better sorbent utilization and other process improvements could reduce CO_2 emissions from the flue gas desulphurization (FGD) unit as well. Table 8-2 shows cost and environmental characteristics for advanced PC plants in comparison with conventional PC technology.

Table 8-2. Cost and Environmental Characteristics for Conventional and Advanced PC Combustion

Technology	Efficiency %	Cost Information (January $1989)			Air Emissions (g/kWh)		
		Capital ($/kW)	Fixed O&M ($/kW-yr)	Variable O&M (mills/kWh)	SO_2	NO_X	CO_2
Conventional PC, 200 MWe FGD	33.4	1,742	38.7	6.4	3.7	3.3	997.3
Advanced PC, 300 MWe advanced FGD	37.6	1,537	29.0	5.2	3.3	3.0	886.9

- **Atmospheric fluidized bed combustion**

In atmospheric fluidized-bed combustion (AFBC), a mixture of solid fuel, granulated limestone sorbent, and inert bed material such as sand or ash are suspended (fluidized) by an upward flow of air. Heat is removed from the combustion zone by producing steam in water-filled tubes passing through the fluidized bed and/or the hot gas stream. A conventional steam turbine utilizes the steam to generate electricity. The suspension provides for better fuel mixing and heat transfer and keeps the combustion temperatures low (about half of a conventional coal plant). This minimizes NO_X formation and provides for near-optimal SO_2 capture by the sorbent. Post-combustion pollution equipment is only needed for particulate control which can be achieved with electrostatic precipitators or fabric filters.

There are two types of AFBC technology: bubbling fluidized bed combustion (BFBC) and circulating fluidized bed combustion (CFBC). The combustion air flow in CFBCs occurs at higher velocities, resulting in a more turbulent bed and entrained flow, whereas in BFBCs the lower air flow velocities create a discernable and measurable bed height. Even though both technologies are commercially available, BFBCs seem to be getting more popular due to increased fuel flexibility, better mixing and heat transfer, higher SO_2 capture, and better combustion efficiency (Sondreal and Jones, 1991). Cost and environmental characteristics are given in Table 8-3.

Table 8-3. Cost and Environmental Characteristics for AFBC

Technology	Efficiency %	Cost Information (January $1989)			Air Emissions (g/kWh)		
		Capital ($/kW)	Fixed O&M ($/kW-yr)	Variable O&M (mills/kWh)	SO_2	NO_X	CO_2
Conventional PC, 200 MWe FGD	33.4	1,742	38.7	6.4	3.7	3.3	997.3
AFBC, bubbl. 200 MWe,	34.3	1,724	34.5	7.6	3.6	1.4	972.9
AFBC, circ. 200 MWe	33.9	1,603	32.4	7.2	3.6	0.9	982.4

- **Pressurized fluidized bed combustion**

Pressurized fluidized bed combustion (PFBC) is operated at a pressure of 6-16 atmospheres. Due to the pressurized conditions and the more efficient steam production, the combustion chamber of a PFBC is generally one-third the size of a conventional furnace. The pressurized gases exiting the combustor are cleaned of particulates, alkali, and other contaminants. The gases are then expanded in a gas turbine to generate electricity and passed through an economizer to preheat the feedwater for the steam turbine cycle before being discharged to the atmosphere. The steam that is generated by the tubes immersed in the fluidized bed is expanded in a conventional steam turbine to produce additional electric power. This combined-cycle (CC) system yields an overall efficiency of up to 39% (EPRI, 1993). PFBCs also come in two different types: bubbling bed (BB) and circulating bed (CB) (see above). Cost and environmental characteristics for PFBC combustion are given in Table 8-4.

Table 8-4. Cost and Environmental Characteristics for PFBC Combustion

Technology	Efficiency %	Cost Information (January $1989)			Air Emissions (g/kWh)		
		Capital ($/kW)	Fixed O&M ($/kW-yr)	Variable O&M (mills/kWh)	SO_2	NO_x	CO_2
Conventional PC, 200 MWe FGD	33.4	1,742	38.7	6.4	3.7	3.3	997.3
PFBC, CC. 340 MWe	38.0	1,508	40.0	6.5	3.2	0.4-0.8	877.1
PFBC, CB 250 MWe	35.2	1,464	30.6	6.5	3.5	0.4-0.8	947.8
PFBC, BB 250 MWe	33.2	1,570	29.7	6.2	3.7	0.4-0.8	1003.9

- **Advanced combustion turbines**

The combustion turbine is a prime mover that converts thermal energy into mechanical work. Inlet air is compressed in the compressor and mixed with fuel in the combustion chamber. The combustion products are then expanded through a fixed nozzle vane and the high-velocity gas stream transfers momentum at constant pressure to the blade, causing tangential rotation that drives an electrical generator. The gas travels through a series of stages (combination of nozzle and blade) in the combustion turbine before exhausting to the atmosphere. A portion of the work produced in the turbine is used to power the air compressor, which is usually mounted on the same shaft as the turbine.

Simple-cycle combustion turbines have efficiencies in the range of 30-35%. However, small (up to 50 MWe), airoderivative turbines are already being marketed with simple-cycle efficiencies of up to 41% (GTW, 1991). In combined cycle operation, the most recent turbines achieve efficiencies of 53-55% (Moore, 1993). By the end of the decade, large (150-250 MWe), heavy-frame machines are expected to operate at efficiencies of 58-61% in combined-cycle mode.

There is significant development work currently underway to design these advanced gas turbines. The advanced designs include (1) steam-injected gas turbines, (2) compressor intercooling, separately, and in combination with steam injection, (3) chemical recuperation of waste heat combined with intercooling and steam injection, and (4) humid-air turbine with an intercooled, regenerative cycle where a saturator adds moisture to the compressor discharge air. Table 8-5 compares technical characteristics of conventional turbines with intercooled steam injected gas turbines (ISTIG).

Table 8-5. Cost and Environmental Characteristics for Advanced Combustion Turbine

Technology	Efficiency %	Cost Information (January $1989)			Air Emissions (g/kWh)		
		Capital ($/kW)	Fixed O&M ($/kW-yr)	Variable O&M (mills/kWh)	SO_2	NO_x	CO_2
Conventional Natural Gas 40 MWe	35.5	362 - 469	0.7 - 1.0	5.4 - 10.3	0.0	0.4	525
Conventional Natural Gas 140 MWe	30.3	342 - 385	0.7	5.4 - 7.6	0.0	1.8	616
ISTIG Natural Gas 50 MWe	47.0	1,142	9.7	9.2	0.0	0.04	397
ISTIG Natural Gas 150 MWe	47.0	893	6.1	5.9	0.0	0.04	397

- **Combined cycle**

  Combined cycle plants consist of a combination of two or more power cycles, each using different working fluids. The most popular combined cycle is the gas and steam turbine combined cycle. In this configuration, a gaseous or liquid fuel is burned to operate a gas turbine to generate electricity. The hot turbine exhaust gases are passed through a steam boiler to produce steam for a steam turbine. The steam turbine utilizes waste heat from the turbine that, in a simple cycle, would have been rejected to the atmosphere. The additional electricity increases the overall system efficiency. Combined cycle generation is ideal for retrofit applications, as either existing combustion turbines or steam turbines can be converted into a combined cycle plant by adding the missing cycle.

- **Integrated gasification combined cycle**

  In the integrated gasification combined cycle (IGCC) technology, coal is first fed to a gasifier where it is partially oxidized to form a raw fuel gas. The raw gas is cleaned to remove sulfur and nitrogen compounds, particulates, and tar. The clean gas is then fired in a gas turbine to generate electricity. The hot exhaust from the gas turbine passes through a waste heat boiler and provides steam for a conventional steam turbine. The steam turbine generates additional electricity.

There are three generic types of gasifiers. The first is the moving-bed process where large particles of coal move slowly downward through the reactor. Countercurrently, a stream of steam and oxygen (or air) moves upward, devolatilizes the coal, resulting in the two products, gas and ash. The second type is the fluidized bed gasifier. In this process, the resultant gas is desulphurized within the reactor using limestone or dolomite as sorbent. Particulates are removed downstream with cyclones. The entrained bed gasifier is the third generic type. Here, the pulverized coal and the oxidant are introduced together and move concurrently with the steam through the gasifier while they react with each other.

IGCC plants are expected to operate at efficiencies of up to 40%. By the end of the decade, IGCC plants are projected to achieve thermal efficiencies of 43% giving them ideal near-term potential for reducing GHG emissions from conventional coal-fired power plants with efficiencies of 35% (Torrens, 1989). Cost and environmental characteristics are given in Table 8-6.

Table 8-6. Cost and Environmental Characteristics for IGCC

Technology	Efficiency %	Cost Information (January $1989)			Air Emissions (g/kWh)		
		Capital ($/kW)	Fixed O&M ($/kW-yr)	Variable O&M (mills/kWh)	SO_2	NO_X	CO_2
Conventional PC, 200 MWe FGD	33.4	1,742	38.7	6.4	3.7	3.3	997.3
IGCC Entrained bed	29.2 - 37.3	1,560 - 1,700	34.0 - 38.4	2.5 - 3.8	0.04 - 1.2	0.2 - 2.5	817 - 1,085
IGCC Moving bed	37.5 - 39.4	1,340 - 1,440	30.2	4.8	0.3 - 3.2	0.2 - 1.8	785 - 862
IGCC Fluidized bed	37.7 - 39.6	1,630 - 1,790	35.2	3.6	1.2 - 3.1	0.9 - 1.4	820

- **Fuel cells**

Fuel cells convert the chemical energy contained in a fuel directly into electricity and heat. Intermediate steps like combustion, conversion of heat to steam, to mechanical energy, and finally to electricity are not needed. Fuel cells can be classified into several categories, with the most important being molten carbonate (MCFC), phosphoric acid (PHFC), and solid oxide (SOFC). Fuel cells are similar to batteries except that the electrodes, that are consumed in a battery, are constantly replenished. Different electrochemical reactions take place, generating electricity, steam, and, depending on the cell type, CO_2. Efficiencies of up to 57% are achievable (see Table 8-7).

Table 8-7. Cost and Environmental Characteristics for Fuel Cells

Technology	Efficiency %	Cost Information (January $1989)			Air Emissions (g/kWh)		
		Capital ($/kW)	Fixed O&M ($/kW-yr)	Variable O&M (mills/kWh)	SO_2	NO_x	CO_2
Conventional PC, 200 MWe FGD	33.4	1,742	38.7	6.4	3.7	3.3	997.3
MCFC, 500 MWe	45 - 47	1,730	27.9	6.2	0.003 - 0.25	trace - 0.08	700 - 755
SOFC, 747 MWe	48.0	1,830	58.5	included in fixed component	0.1	0.11	611 - 682

- **Combined heat and power systems - cogeneration**

A combined heat and power (CHP) system consists of an engine or a turbine to drive an alternator and a heat recovery system. In addition to the electrical power generated by the alternator, some or all of the heat from the prime mover can be used in the form of steam, hot water, or hot gases. Combined heat and power systems can reach thermal efficiencies of up to 80% and more. Small CHP systems convert about 20% of the fuel input into electric power and about 55-60% into useful heat. In larger systems, the electric power output may be as high as 40% of the fuel input. Potential applications are district heating and cooling and process heat for industrial purposes. The very high system efficiencies of CHP systems bear enormous potential for fuel savings and resulting GHG mitigation (see Table 8-8).

Table 8-8. Cost and Environmental Characteristics for CHP Systems

Technology	Efficiency %	Cost Information (January $1989)			Air Emissions (g/kWh)		
		Capital ($/kW)	Fixed O&M ($/kW-yr)	Variable O&M (mills/kWh)	SO_2	NO_x	CO_2
Conventional PC, 200 MWe FGD	33.4	1,742	38.7	6.4	3.7	3.3	997.3
CHP, 200 MWe	80.0	na	na	na	1.5	several criteria 1.4	416.8

In back-pressure CHP plants, the steam leaving the steam turbine is condensed at conditions that generate hot water which can be directly fed into a district heating system. Heat and electric power are produced at a fixed ratio. In extraction-type CHP plants, steam is extracted from the steam turbine to heat the water for the district heating system. This system is more flexible as it allows for a variable ratio of heat and power generation. The higher the heat demand, the more steam needs to be extracted and the lower the electric power generation.

- **Improvements in transmission and distribution systems**

  In industrial countries, power transmission and distribution (T&D) systems generally have losses that are in the range of 5-10%. In developing countries and countries with economies in transition, this number can be substantially higher. A survey of the Asian Development Bank (ADB) among its member countries showed T&D system losses ranging from 15% to 37% with an average net system loss of 21.5% (Burrel, 1991). The losses may be associated with lack of financial resources to expand and maintain the systems, chronically overloaded systems, inadequate billing and collection infrastructure, and theft.

  Reducing T&D losses provides, in essence, pollution-free power mostly at very favorable financial terms. Loss reductions can be achieved through system rehabilitation projects, use of capacitators and synchronous condensers to correct system power factors, rigorous loss reduction programs, improved billing and collection procedures, and increased security and vigilance. An ADB-financed project in Pakistan revealed that rehabilitating the distribution system can provide power at an investment cost of US$135 per kW. Power factor improvements have been shown to generate additional power at costs of US$100-200 per kW (Burrel, 1991).

8.2.2 Power Sector: Fuel Switching and Repowering

Fuel switching in the power sector is an option to reduce GHG emissions. In these cases, coal would be substituted for low- or zero-carbon fuels, namely natural gas and nuclear (hydro and other renewables are covered in Chapter 9). Natural gas emits about half the CO_2 per GJ of fuel and in many applications can replace oil or coal directly at relatively low cost. Nuclear power generation is assumed to have no GHG emissions although the fuel-processing cycle uses an extensive amount of energy, which may result in GHG emissions.

There are a number of uncertainties associated with nuclear power. Public acceptability, waste management, and financing availability are only a few of the issues that will affect nuclear power's future role. In addition, there is considerable debate about the cost of nuclear power. Cost items like decontamination and decommissioning, long-term waste management, and indirect cost for emergency evacuation planning are usually not reflected in cost assumptions of nuclear power. Including these cost items may significantly change the economic competitiveness of an expanded nuclear program.

Repowering is a somewhat different approach. Based on criteria like age, size, and control equipment, a pool of coal-fired power stations can be drawn up that may be suitable for repowering. In the repowered unit, the old firing technology is replaced with any of the new, advanced technologies.

8.2.3 Power Sector: Post-Combustion Options

The discussion of post-combustion options has, to this point, mostly centered around the mitigation of CO_2 emissions. Recent research and development work has identified a variety of technologies in this category, some of which have been derived from industrial applications. At this time, all post-combustion technologies are expensive and energy-intensive. This is due mostly to low CO_2 concentrations in the flue gas, the presence of corrosive elements, the dominance of N_2 in the flue gas that, like CO_2, is relatively inert, etc. The most promising technology for CO_2 capture from power plants appears to be the chemical stripping method. The technologies are discussed further in DOE (1993).

8.2.4 Fuel Supply: Extraction and Delivery System

- **Natural gas supply system**

Typical components of a natural gas supply system include production wells, gas-processing plants, transmission pipelines, storage and injection/withdrawal facilities, and distribution systems. The system incurs losses at each of these steps, and as natural gas consists primarily of methane (a greenhouse gas), these losses contribute to GHG emissions.

As shown in Table 8-9, there is a variety of emission reduction options currently available to the natural gas supply sector. In addition, development work is in process for new technologies and practices to reduce methane emissions at every stage of the supply system. Technologies that may play a stronger role in the near future include (1) installation of catalytic converters on reciprocating engines, (2) use of "smart" regulators in distribution networks, (3) use of metallic coated seals, (4) use of sealant and cleaner injections in valves, and (5) use of composite wraps for pipeline repair.

The U.S. Environmental Protection Agency (EPA) found that the technically feasible methane emission reduction potential of the natural gas system may be as high as 33% (in 2010) using available technologies. The cost-effective reduction potential is estimated to be about 25% (in 2010) (EPA, 1993).

- **Coal mining**

During the coalification process, gases are produced and entrapped in the coal. When the coal is mined, the gas is released to the atmosphere. The main component of the coal bed gas is methane (CH_4) (>80%) with the remainder being CO_2, N_2, O_2, H_2, and He. The heating value is comparable to natural gas. Coal rank, pressure (depth), and temperature determine the amount of gas in the coal. For the U.S., average CH_4 release factors have been determined by coal type (DOE, 1990) with the following results: 22 m^3/tonne (anthracite), 7 m^3/tonne (bituminous/subbituminous), and 0.3 m^3/tonne (lignite).

Standard procedure in a coal mine is to collect and vent the CH_4 released into the mine shaft. Possible GHG mitigation first seeks to optimize the CH_4 release by short-, medium-, and long-hole drilling for pre- and post-drainage (Lama, 1991), and drilling of cased wells with perforations. The coal bed CH_4 is collected and utilized for on-site electric power generation. The CH_4 recovery rate of this system is estimated to be 50% (EPA, 1990). Capital costs are $690,000 per well with an annual capacity of 2,000,000 m^3 at operating costs of $80,000 per year (Kuuskraa, 1989).

Table 8-9. GHG Mitigation Options for the Natural Gas Supply Sector

Stage of Supply System	Emission Source	Mitigation Option	Comments
Production and Processing	Pneumatic devices (used on heaters, separators, gas hydrators, gathering pipelines)	Replace high-bleeding pneumatics with no- or low-bleeding devices	U.S. emission reduction potential: 7.1% Very cost-effective
	Gas dehydrators (remove H_2O from gas stream with glycol, emissions occur during glycol regeneration)	Install flash tank separator, use recovered methane as fuel in glycol regeneration unit	U.S. emission reduction potential: 7.1% Generally cost-effective
	Fugitive emissions (through leaks at damaged seals or corroded pipeline)	Implement inspection and maintenance programs	U.S. emission reduction potential: 5.7% Not cost-effective
Transmission	Pneumatic devices	Replace high-bleeding pneumatics with no- or low-bleeding devices	U.S. emission reduction potential: 3.4% Very cost-effective
	Reciprocating engines (drive compressor engines)	Use turbine engines for compression when constructing new pipeline or retiring reciprocating engines on existing pipelines	U.S. emission reduction potential: 3.7% Other operational factors must be considered when selecting compressor engines, site-by-site decision
	Venting during routine maintenance for repairs	Use portable evacuation compressors to pump gas to an adjoining section	U.S. emission reduction potential: 0.6% Cost-effective in Canada Not cost-effective in U.S.
	Fugitive emissions (through inadequately sealed valves, fittings, assemblies, or corroded pipeline	Implement inspection and maintenance programs at compressor stations	U.S. emission reduction potential: 7.1% Cost-effective
Distribution	Fugitive emissions from gate stations	Implement inspection and maintenance programs	U.S. emission reduction potential: 3.1% Cost-effective
	Fugitive emissions from subsurface piping	Replace leaking pipe or joint or insert repair material in old pipe	Very expensive Not cost-effective

8.2.5 Screening Technology Options for Analysis

A first step in conducting an analysis is to screen the available options and select a set for detailed evaluation. This screening will eliminate expending limited resources on options that will have little effect or that may not be practical for consideration in a country.

The method for screening and selecting options for analysis relies on the use of expert judgement in the consideration of a country's energy situation. It is primarily a qualitative evaluation of which options may offer the best possibilities. The basic steps that are used in the screening are discussed in the following sections.

8.2.5.1 Identify major GHG sources - current, future

First, major sources of GHG emissions from the current energy supply system need to be identified based on the inventory of GHG emissions. Mitigation options that operate on all the major sources should be considered for inclusion in the analysis.

In reviewing the information in a GHG inventory, attention should be focused not only on large individual sources (e.g., power stations, coal mines) but also on small, distributed systems (e.g., natural gas distribution lines) that may have small individual emissions but which may, in aggregate, be a major contributor.

In addition to considering current GHG emission sources, the future configuration of the energy supply system needs to be evaluated. Future major sources of GHG emissions need to be identified. Given that the planning horizon for climate change analyses may extend over several decades, the configuration of the energy supply system is likely to change significantly. Many of the existing facilities may be retired during the period. New facilities of the same type or of very different types may be brought on-line. The candidate facilities for the future energy supply system need to be included in the list of possible major GHG emission sources.

For each of the GHG sources identified, a list of mitigation options should be tabulated. This list should include options that reduce emissions from individual sources (e.g., improving combustion efficiency, fuel switching) and options that are applied on a system level (e.g., modified dispatching of electric power plants). The mitigation options identified above can be used as an initial list of possibilities.

8.2.5.2 Evaluate candidate options

Criteria that can be used to evaluate candidate options are listed in Chapter 2, Table 2-1. Here several criteria are discussed which are important for conventional supply options.

- **Applicability**

  Options that cannot be realistically considered to apply in a country's unique circumstances can be dropped from further analysis. For example, it is not reasonable to include the use of natural gas in a country that has no ready access to gas supplies either domestically or from imports. As another example, it may not be reasonable to consider a repowering program for power stations that are very old and soon to be retired.

- **Potential Effectiveness**

  The potential for a given mitigation option to have a significant impact on GHG emissions is another consideration used for screening. Options that will have only a small impact may be relegated to a lower priority when the analysis is carried out. The focus should be on those options that have the largest potential for effectiveness.

- **Potential Cost**

  Options may be screened on the basis of the potential cost to implement. While actual costs would have to be determined during the course of the analysis, there is generally enough information to identify which options would be prohibitively expensive to carry out.

- **Potential for Other Environmental and Social Benefits and Costs**

  Some options may provide other environmental and/or social benefits in addition to the reduction of GHG emissions (e.g., improved system efficiency can reduce the import fuel bill and the production of other air pollutants). These additional benefits can warrant inclusion of an option for study. Also, some options may have other environmental and/or social costs (e.g., a hydropower station that floods valuable land or displaces large groups of people) that exclude them from consideration.

- **Implementation Barriers**

  Some candidate options may be screened out on the basis of barriers (e.g., policy, regulatory, public concern) that would be very difficult to overcome. This would eliminate from the analysis those options that, while possibly effective, probably could not be implemented.

8.2.5.3 Select options for analysis

A set of mitigation options that will be subjected to more detailed analysis should be drawn up from the above considerations. This list will help focus the analysis efforts in the most productive directions.

In developing this list, several issues should be kept in mind. First, mitigation options can be considered for application both individually and in combination with other options. For example, an individual option would be to improve the efficiency of operation of electric power stations. A combined set of options would be to improve efficiency and switch from coal to natural gas. In some cases, a particular mitigation option may not be particularly effective by itself but may be very effective when taken in combination with other options.

Second, the mitigation options in the conventional energy supply system must be considered for their interactions with other sectors. What is done in one sector may have substantial effects, positive or negative, in other sectors.

Third, the analysis of mitigation options should be carried out on an iterative basis. Some options that have been selected for analysis may, on more detailed evaluation, prove to be ineffective or inappropriate. Likewise, some options that were initially screened out may, after initial results are completed, need to be reinvestigated for their potential applicability.

8.3 ENERGY RESOURCE ASSESSMENT

The investigation of the availability of energy resources determines what energy supplies are currently available and might be available in the future. The principal resources of concern here are fossil fuels: oil, natural gas, and coal. Resources other than fossil fuels (e.g., renewable energy resources) are also an important part of a resource analysis, and are covered in Chapter 9.

The resource analysis of fossil fuels is a geological determination of the location of resources, the quantity of resource available, and the economics of extraction. One method of categorizing fossil fuel resources is the procedure used by the U.S. Geological Survey. Resources are categorized into the following groups:

- Identified Resources
  - -Measured
  - -Indicated
  - -Inferred
  - -Economically recoverable
  - -Marginally economic
  - -Subeconomic

- Undiscovered Resources

When tabulating a country's fossil fuel availability, the term "reserves" is applied only to the measured and indicated resources that are economically recoverable. This gives a much more realistic picture of the availability of commercially exploitable energy.

In addition to identifying domestic resources, the resource analysis needs to determine the opportunities for importing fossil fuels. Oil, natural gas, and coal are all internationally traded commodities that can contribute to a country's supply system.

The significance of the resource analysis on the analysis of GHG mitigation options is two-fold. First, the availability of domestic fossil energy resources will determine what possibilities a country has to switch to lower carbon-content fuels. For example, a country that has large domestic coal reserves but limited natural gas will have difficulty in substituting for the coal in power generation. Second, the exploitation of fossil fuel resources (e.g., coal mining, crude oil extraction, natural gas extraction) is itself a source of GHG emissions that may require mitigation measures.

8.4 ENERGY TECHNOLOGY CHARACTERIZATION

Energy technologies that utilize the available energy resources need to be identified and characterized. A consistent and comprehensive data set for each candidate technology needs to be assembled. The technologies under consideration must cover the entire energy system, both in its current configuration and in possible future configurations.

Table 8-10 gives a typical list of energy supply system technologies that could be included. The list is intended to be illustrative and not comprehensive.

All steps of the fuel cycle should be included even though some are not significant from the standpoint of GHG emissions. Some pieces of the system, although not major contributors to GHG emissions, will affect how the system is designed. For example, the use of natural gas as a substitute for coal in power generation will be constrained by the availability of pipelines to bring the gas to power station sites.

Table 8-10. Typical Conventional Energy Supply Technologies

	FUEL CYCLE STEP EXTRACTION	PROCESSING	TRANSPORT	CONVERSION
OIL	On-shore Off-shore Primary recovery Secondary recovery Tertiary recovery	Refining	Pipeline Tanker Barge Truck Rail	Combustion
NATURAL GAS	On-shore Off-shore Associated Non-associated Primary recovery Secondary recovery Tertiary recovery	Well-head processing Liquefaction	Pipeline Rail (LNG) Ship (LNG)	Combustion
COAL	Underground Surface	Cleaning Solvent refining Gasification Liquefaction	Rail Barge, ship Truck Slurry pipeline	Combustion
NUCLEAR MATERIALS	Mining	Conversion Enrichment Fuel fabrication	Rail Truck	Reactor fuel

The characterization of these technologies at this stage in the analysis process consists of assembling information for each technology in a consistent fashion. The information to be gathered is illustrated on Table 8-11. Sources of this information include IPCC (1993), and also EPRI (1993).

Table 8-11. Typical Energy Supply Technology Data Requirements

PARAMETER	EXAMPLES
Engineering Performance Data	
Energy output	Type of products output Range of output capacity
Energy input	Input fuel Input materials Restrictions on inputs
Thermodynamic efficiency	Current Future
Performance limits	Capacity design, maximum, minimum Operational limitations
Technology status	Commercially available Research Pilot plant
Economic Data	
Capital cost	Labor, materials Interest during construction Foreign and domestic component
Non-fuel operating cost	Labor, materials Taxes
Financial data	Interest rate Return on investment Discount rate Foreign exchange
Environmental Data	
GHG emission rates	CO_2, Methane, Others Without mitigation options With mitigation options
Air pollutant emission rates	SO_x NO_x Particulates Others
Other environmental burdens	Water pollution Solid waste generation Land use Occupational health and safety

8.5 CONSTRUCTING SCENARIOS

The construction of a baseline scenario for energy supply should be conducted as part of demand/supply balancing in an energy system model. Similarly, analysis of mitigation options for conventional energy supply should be conducted in an integrated framework. This integrated analysis will identify those options affecting conventional energy supply that appear attractive. The analyst must then consider barriers to the implementation of these options and identify policies to promote their adoption.

8.6 MITIGATION POLICIES

The steps that need to be taken to encourage adoption of mitigation technology options must be identified and evaluated as part of the overall analysis. Tables 8-12 to 8-15 list several categories of implementation strategies with examples. These are adapted from DOE (1989) and are only illustrative of the many possibilities. It is not possible to select a set of implementation strategies that will be applicable to all countries. The unique situations of each country require a careful evaluation of alternative implementation techniques and the selection of the ones most suited to local conditions.

Each of the implementation options has different issues associated with its use. For example, regulations tend to be rigid and not easy to adjust to changing circumstances. They also require a large effort to enforce. Fiscal measures are more flexible, but it is not as easy to determine what level of fiscal measure (e.g., tax or subsidy) is required to produce the desired alterations to the energy system. Information measures are useful, but it is difficult to judge how effective they are. Research and development activities will generally require longer time periods to produce results. In reality, countries will need to employ a combination of implementation measures that match local conditions.

Integrated Resource Planning (IRP) is an approach that can be helpful in evaluating and encouraging adoption of mitigation options for the electric sector. Governments can encourage or require utilities to utilize IRP techniques, which provide a framework for considering demand- and supply-side options in an integrated manner.

Table 8-12. GHG Mitigation Implementation Options - Fiscal Incentives

OPTION	DESCRIPTION	EXAMPLES
Emission Fees	Fees placed on emissions of GHGs	CO_2 emission fee on energy facilities
Tradeable Emission Rights	Requiring the acquisition of rights to emit GHGs by trading with other sources; total emission quantities to be government-controlled	Reforestation project to accompany new energy facility construction Retirement of one facility before a new one can be built
Deposit-Refund System	Deposits taken on products with high GHG content or with high GHG sources used in manufacture; refunded upon proper disposal	Deposit on CFC-containing equipment
Taxes - taxes levied on GHG sources		
Excise taxes	Taxes levied on specific products	Tax on electricity generated by coal Tax on oil production
Taxes on firms	Taxes levied on companies	Tax on coal companies Tax on oil companies
Taxes on individuals	Personal income taxes	Tax credits for use of efficient equipment
Property taxes	Taxes on land and facilities	Taxes on high GHG-emitting facilities Tax credits for low GHG-emitting facilities
Tariffs	Taxes levied on imported goods and services	Taxes on imported, high carbon-content fuels
Subsidies - rebates or lower tax rates to low-GHG-emitting sources		
Direct Government Expenditures - government funding for activities		
Research and development support	Government funding of R&D programs	Energy efficient technology research GHG monitoring research
Direct government purchases	Government purchase of low GHG-emitting technology	Purchase for government use Purchase for private use

Table 8-13. GHG Mitigation Implementation Options - Regulatory Measures

OPTION	DESCRIPTION	EXAMPLES
Controls - measures that are direct controls on the development and operation of the energy system		
Bans	Strict prohibition of the use of certain technologies	Ban on new coal-fired equipment Ban on flaring gas from production wells
Emission controls	Limits on the emissions of GHG from plants in the energy system	GHG emission quotas on power plants
Input controls	Limits on the inputs to energy system facilities	Restriction on quantity of coal consumed
Consumption controls	Limits on the consumption of energy or energy-related services	Electricity use restrictions Restrictions on energy-consuming activities
Price controls	Limits on prices for energy - price ceilings/floors	Price floors on carbon-based fuels Price ceilings on non-carbon or low-carbon fuels
Rate-of-return regulation	Requirements on the rate-of-return calculation process for energy supply companies	Modification of rate base for electric utilities to allow for lower-carbon but more expensive systems
Standards - measures that set performance standards on energy supply systems		
Technology standards	Performance, process, equipment and design, and product standards for energy supply technologies	Requiring the use of specific electricity generation technologies Efficiency standards for power plants
Licensing and certification	Conditions placed on energy supply operations that must be met before they can operate	Licensing of power plant designs Site permits

Table 8-14. GHG Mitigation Implementation Options - Information Strategies

OPTION	DESCRIPTION	EXAMPLES
Advertising and Labeling	Requiring GHG information on equipment	GHG emission labels on energy supply system equipment
Education	Information dissemination programs	Information programs for utility planners Information programs for the general public
Moral Suasion	Adversarial information program to persuade energy system operators	Program to promote low GHG technologies under threat of regulation
Signaling	Conveying signals to energy system operators about desired actions	Advanced notices of proposed rules Non-mandatory standards

Table 8-15. GHG Mitigation Implementation Options - Research and Development

OPTION	DESCRIPTION	EXAMPLES
Public Invention Support Programs	Providing support to commercialization of innovative technologies	Energy-efficient technology awards
Commercialization Education	Dissemination of information on commercializing innovative technologies	Information programs for private sector entrepreneurs
Provision of Specialized Information	Information program for energy operators to accelerate technology change	Extension programs
Demonstrations	Full-scale demonstrations of innovative technology	Demonstration of new electricity generating technology

REFERENCES

Burrel, A. D. 1991. Coal, The Environment and Development: Technologies to Reduce Greenhouse Gas Emissions. In: *Proceedings of International Conference on Coal, The Environment and Development: Technologies to Reduce Greenhouse Gas Emissions*, Sidney, Australia, 18-21 November 1991.

DOE. 1993. *The Capture, Utilization and Disposal of Carbon Dioxide from Fossil Fuel-Fired Power Plants. Volumes 1 and 2.* DOE/ER-30194. Washington, DC. July.

DOE. 1990. *An Evaluation of the Relationship between the Production and Use of Energy and Atmospheric Methane Emissions.* DOE/NBB-00888P. Richland, VA.

DOE. 1989, *A Compendium of Options for Government Policy to Encourage Private Sector Responses to Potential Climate Changes*, 2 volumes, U.S. Department of Energy, Office of Environmental Analysis, Report No. DOE/EH-0103.

EPA. 1990. *Preliminary Technology Cost Estimates of Measures Available to Reduce Greenhouse Gas Emissions by 2010.* Prepared by ICF Resources. Washington, DC.

EPA. 1992. *Compilation of Air Pollutant Emission Factors*, U.S. Environmental Protection Agency, Report AP-42, Research Triangle Park, NC.

EPA. 1993. *Anthropogenic Methane Emissions in the United States.* Report to the Congress. Prepared by the Global Change Division. Office of Air and Radiation. Washington, DC.

Electric Power Research Institute (EPRI). 1993. *TAG - Technical Assessment Guide, Volume 1: Rev. 7: Electricity Supply - 1993.* EPRI P-6587-L. Palo Alto, CA.

Gas Turbine World (GTW). 1991. *The 1991 Handbook, Volume 13.* Pequot Publishing. Southport, CT.

IPCC. 1993. *The IPCC Technology Characterization Inventory, Phase II Report, Volume 1.* Prepared for the Intergovernmental Panel on Climate Change.

Kuuskraa, Velo, *et al.* 1989. Steps to Assess Resource Economics Covered. *Oil and Gas Journal.* December 25, 1989. pp 121-125.

Lama, Ripu D. 1991. Methane Gas Emissions from Coal Mining in Australia: Estimates and Control Strategies. In: *Proceedings of International Conference on Coal, The Environment and Development: Technologies to Reduce Greenhouse Gas Emissions*, Sidney, Australia, 18-21 November 1991.

Sondreal, E.A., and M.L. Jones. 1991. *Clean Coal Technologies: Low Rank Coal Perspective.* Paper presented at "The Role of Clean Coal Technologies - An International Seminar", Argonne, IL (June 27).

Torrens, Ian M. 1989. "Greenhouse Effect" section in Special Report "Meeting Highlights Global Power-Technology Issues". *Power.* 133(12): 15-28

CHAPTER 9
RENEWABLE ENERGY SUPPLY

9.1 INTRODUCTION

Renewable energy can be defined initially as any energy source that is derived directly or indirectly from solar energy. In the broadest sense, however, almost all of the energy we use today, including fossil fuels, can be considered a form of solar energy. The most familiar forms of energy, such as wood, oil, gas, and coal, are embodied forms of solar energy gathered, stored, and transformed by natural processes.

Climate change due to emissions of GHGs, particularly CO_2, becomes an issue when stored solar energy is converted to useable forms of energy (heat, electricity, fuels, chemicals) at a rate far exceeding the rate of formation. For coal, oil, and natural gas, the ratio of time between formation and use is on the order of 1 million to one: that is, the world uses in one year what took natural processes one million years to create. Only biomass among these stored forms has a time ratio that is within a human time frame of years or decades. Renewable energy can now be defined as forms of solar energy that are available and replenished in time scales no longer than human lifetimes.

Given this definition of renewable energy, it becomes clearer why renewable energy is an important option for mitigating climate change. Because renewable energy creates little if any net greenhouse gas emissions, its use will not disrupt the radiative energy balance of the earth's atmosphere and will permit sustainable, long-term mitigation of climate change. The renewable energy option will allow climate change mitigation, energy use, and economic development to proceed in synergy rather than in opposition.

The remainder of this chapter will discuss what information, data, and analytic tools are needed to identify, screen, and characterize renewable energy options. The information and data needed include:

- economic and social development goals and needs
- energy end uses and tasks to be performed
- characteristics of energy needs
  - scale (total requirements, grid/off-grid, centralized/distributed, etc.)
  - timing of energy needs (duration, seasonality, diurnal, etc.)
- available energy resources
- technology characterization

Once energy needs are defined within the larger context of economic and social development needs and plans, renewable energy resources and technologies can be identified and evaluated for incorporation into this larger context. The analytical tools needed include systematic methods to inventory renewable energy resources and to evaluate the most appropriate applications of these resources.

This chapter presents an overview of renewable energy options and discusses resource assessment and characterization of renewable energy technologies. Methods for analyzing renewable energy options within an integrated framework are discussed in Chapter 3. Policy options for encouraging adoption of renewable energy technologies are briefly described at the end of this chapter.

9.2 MITIGATION TECHNOLOGY OPTIONS

Renewable energy supplies encompass a broad range of resources, and numerous technologies can be used to tap those resources. Table 9-1 lists the major technologies and the following discussion briefly describes each technology and its applications. Although many of these technologies are still under development, most have entered commercial markets around the world at some level. Some, such as hydropower and biomass technologies, have achieved sizeable market penetration, while others (e.g.,

Table 9-1. Renewable Energy Technologies

RESOURCE	TECHNOLOGY	END-USE APPLICATION			
		Electricity[a]	Industry	Buildings	Transport
Solar	Photovoltaics - Flat Plate	✓			
	Photovoltaics - Concentrator	✓			
	Solar Thermal Parabolic	✓	✓		
	Solar Thermal Dish/Stirling	✓			
	Solar Thermal Central	✓	✓		
	Solar Ponds	✓	✓	✓	
	Passive Heating			✓	
	Active Heating			✓	
	Daylighting			✓	
Wind	Horizontal Axis Turbine	✓			
	Vertical Axis Turbine	✓			
Biomass	Direct Combustion	✓	✓	✓	
	Gasification/Pyrolysis	✓	✓	✓	✓
	Anaerobic Digestion	✓	✓	✓	
	Fermentation				✓
Geothermal	Dry Steam	✓			
	Flash Steam	✓			
	Binary Cycle	✓			
	Heat Pump			✓	
	Direct Use		✓	✓	
Hydropower	Conventional	✓			
	Pumped Storage	✓			
	Micro-hydro	✓			
Ocean	Tidal Energy	✓			
	Thermal Energy Conversion				

[a] Electricity generated by any of these methods can be used in the remaining end-use applications either to meet power demands directly (e.g., industrial electricity inputs, buildings lighting demand, or electric vehicles) or as an input to end-use fuel production (e.g., hydrogen produced via electrolysis).

photovoltaics) are used in important but relatively limited applications today. Research and development activities continue to improve all of these systems to enhance their ability to meet future energy requirements, and new systems that are still in the early stages of development may provide additional opportunities.

9.2.1 Solar Energy

Solar technologies use the sun's energy directly to generate energy for industrial processes, buildings, and transportation as well as electricity for general consumption in all three of these end-use sectors. Given the large size of the solar resource, these technologies are not constrained by feedstock requirements but rather by costs and "institutional" obstacles such as performance (e.g., intermittent operation), perceived risks, and siting issues.

- **Photovoltaics**

  Photovoltaic (PV) devices convert the energy contained in sunlight directly into electricity-using modules composed of multiple PV cells. Two broad categories of PV devices exist: flat-plate and concentrating. Concentrator systems uses lenses to focus radiation onto just a few, highly efficient PV cells and only use direct beam sunlight, while flat-plates utilize the whole of the incident solar radiation, including diffuse (scattered) and direct insolation.

  Today's annual PV market is only about 50 MW worldwide, but market growth during the last few years has been 25% per year. PV systems are currently cost-effective in some consumer products (e.g., watches and calculators) and distributed and remote power generation (e.g., village power). For example, over 2000 small residential PV systems have been installed in the Dominican Republic under a unique revolving credit system that permits rural clients to borrow funds to purchase these systems and pay the loans back as they save money from avoided kerosene purchases. Similarly, 15,000-20,000 systems have been installed in Mexico under the government's rural development program. As costs continue to decline in the next 5-15 years, opportunities for PV systems will expand, allowing them to compete with large-scale conventional power generation in the next century.

- **Solar Thermal - Electric**

  Solar thermal technologies collect the sun's radiant energy to create a high-temperature heat source that can be converted into electricity via a number of thermodynamic conversion cycles. Parabolic trough technologies employ a field of parabolically-shaped solar collectors that focus the sun's energy onto specially-coated metal pipes surrounded by glass tubes containing a heat transfer fluid (such as synthetic oil). Parabolic dish systems use a modular mirror system that approximates a parabola and creates a high energy flux at the focal point where an external combustion Stirling engine converts the heat into electricity. Central receivers use a large field of sun-tracking mirrors (heliostats) that reflect the incident radiation onto a tower-mounted thermal receiver. Finally, solar pond systems collect and store solar energy in a liquid medium (usually a large basin of water with a salt gradient to suppress heat loss), which can then be converted to electricity using a closed Rankine Cycle engine.

  Solar thermal technologies are currently in the development/demonstration phase, with projects designed to prove the reliability and operation of such facilities in numerous locations worldwide. Many systems employ thermal storage devices or energy backup (so-called "hybrid" systems) to overcome issues associated with the intermittent nature of the solar resource (e.g., power generation during cloudy weather). Likely applications for these technologies will be village power (especially the parabolic dish/Stirling systems) and centralized electricity generation in the late 1990s and early 21st century.

- **Solar Thermal - Industrial Process Heat**

  Solar thermal technologies for industrial process heat (IPH) utilize technologies and principles similar to solar thermal electric technologies in generating a high- or medium-temperature heat source. The heat generated from these systems can then be used to supply energy for general industrial processing needs or for specialty processes, such as the detoxification of hazardous wastes.

  Development and deployment opportunities for these technologies are closely tied to those for solar thermal electric systems since they employ similar thermodynamic and physical principles.

- **Solar Building Technologies**

  Solar building technologies include active and passive heating and cooling systems, as well as daylighting. Today, there are more than two million solar water heaters installed in Japan and 600,000 in Israel; significant numbers are also found in many other countries, notably the United States, Kenya, China, Turkey, and Papua New Guinea. Active heating systems provide hot water and space heating for residential or commercial buildings utilizing a collector that receives or absorbs the incident solar energy and transfers it to a working fluid (water, oil, or air) for direct use or storage. Active solar cooling technologies include solar desiccant systems that use a drying agent to adsorb water vapor in building circulation air; solar heat is then used to dry or regenerate the desiccant for re-use. Another cooling technology, the solar absorption system, is based on traditional refrigeration technologies but uses solar heat to provide much of the energy, although some mechanical assistance is typically required.

  Passive heating and cooling systems use little or no mechanical assistance, relying rather on the design of the building to achieve specific thermal requirement goals. Passive space heating uses natural heat transfer processes to collect, store, and distribute heat. Techniques in practice today include direct gain systems (e.g., south-facing windows), thermal storage walls, attached sunspaces (e.g., greenhouses), roof storage using water that collects heat and is distributed via convection, and convective loops based on a thermosiphon principle common in solar hot water heaters but using air as the working fluid.

  Conventional passive cooling techniques maximize natural ventilation, incorporate well-insulated and low-emissivity building materials, and utilize advantageous landscaping methods (e.g., shade tree planting). More advanced practices involve earth cooling tubes that use lower underground soil temperature to cool incoming air, evaporative cooling using water, and ice ponds (in which ice from the winter months is used to cool circulating air as it melts).

  Finally, daylighting simply involves the effective use of natural light to provide illumination. This is primarily achieved through building design although advanced optical switching materials and low-emissivity coatings for windows can further the effective use of daylighting.

9.2.2 Wind Energy

Wind technologies convert the energy of moving air masses at the earth's surface to rotating shaft power that can be directly used for mechanical energy needs (e.g., milling or water pumping) or converted to electric power in a generator. Two major types of turbines exist and are defined based on the axis of blade rotation: horizontal-axis (which currently dominate commercial markets) and vertical-axis turbines.

Wind energy has proven the most cost-competitive renewable electricity technology for the bulk power market to date; however, its use is also very well-suited to remote and distributed applications. Hybrid applications, in which a wind turbine is coupled with another renewable energy source (e.g., PV) and/or a conventional back-up unit (e.g., diesel generator), are attracting much interest in the remote power market. For example, the fishing village of Xcalac in Mexico uses a hybrid system composed of six 10 kW wind turbines, an 11.2 kW PV array, and a diesel backup generator to provide 100% of its power.

As of 1994, there were over 1700 MW of installed wind turbines in the world, the majority of which are located in California in the U.S. Most of these are in the 50-150 kW size range and are providing power to the electric utilities in that state. Newer turbines being installed today for bulk power are closer to the 150-300 kW range, and systems in the future are expected to be much larger, reaching sizes over 500 kW.

9.2.3 Biomass Energy

Biomass energy is a term that includes all energy materials derived from biological sources, including wood wastes, agricultural residues, food industry wastes, sewage, municipal solid waste (MSW), and dedicated herbaceous or woody energy crops. The potential size of the biomass resource is quite large on a global scale, and the ability to utilize existing residue streams that may provide low-cost feedstocks offers attractive near-term opportunities for biomass use. In the longer term, the development of sustainable, dedicated biomass energy plantations may further expand the resource base and help reduce the costs of energy produced from biomass.

- **Biomass Electric Technologies**

  Cogeneration of biomass in the industrial sector provides the largest share of biomass-derived electricity today. Low-cost feedstocks are often critical to the economic viability of biomass energy use in these markets, and thus waste materials such as agricultural and forest products residues, food processing wastes, and MSW can be attractive feedstocks for direct combustion in boiler systems. Dedicated electricity generation also benefits from the use of waste materials although the development of dedicated biomass energy crops may provide low-cost feedstocks in the long term while simultaneously extending the resource base.

  The energy, environmental, and economic performance of biomass electric technologies can also be enhanced by moving to advanced conversion technologies, such as gasification/gas turbine technologies. Gasifiers convert biomass to a synthetic gasoline (syngas) containing chiefly CO and hydrogen, which can then be combusted in an aeroderivative gas turbine. The theoretical efficiency of such a process is much higher than that of the conventional direct combustion/steam turbine process, thereby reducing feedstock requirements and enhancing the economics of power generation.

  Biomass energy systems are emerging as an economically viable option for satisfying power needs in both industrialized and developing countries (on-grid and off-grid). Where the resource conditions are favorable, power systems based on feedstocks grown on short-rotation forestry (SRF) plantations can produce electricity at costs that are comparable with conventional fossil-based alternatives (see Carpentieri *et al.*, 1993; Russell *et al.*, 1992; and Perlack *et al.*, 1991). Potential SRF biomass-to-electricity projects include:

- the conversion of agricultural processing facilities (e.g., sugar mills, rice mills, sawmills) to cogenerate heat and power for on-site processing needs and export of excess power to the local distribution grid using a combination of mill wastes and plantation grown biomass;

- the conversion of fossil-fired electric generation facilities to burn or co-fire plantation-grown biomass; and

- and the development of stand-alone biomass-based power generation relying exclusively on SRF plantation-grown feedstocks.

The interest in biomass-to-electricity projects comes not only from its potential as a low-cost supply of power but for its potential to mitigate a host of environmental concerns such as reducing the rate of CO_2 buildup in the atmosphere by sequestering carbon, by substituting for fossil fuels, and by replacing wood use from existing forests. When SRF feedstocks are grown renewably, biomass contributes no net buildup of atmospheric carbon because the carbon released during burning is extracted from the atmosphere during photosynthesis.

Direct combustion technologies can continue to improve and expand their market share in industrial cogeneration and utility markets, and the advanced power cycles discussed above are likely to move out of the demonstration phase and into commercial operation near the turn of the century.

- **Biomass Heating Technologies**

The most common use of biomass is direct combustion for residential space heating and cooking and for industrial process heating. In the residential sector, improved wood cooking and heating stoves can offer advantages in terms of reduced fuel requirements (thus, lower costs or less time spent collecting biomass) and improved emissions characteristics. In industrial processes (e.g., the sugar cane industry or pulp and paper industry), biomass wastes can be used more effectively by increasing boiler efficiencies and enhancing residue collection activities.

- **Biomass Transportation Fuels**

Fuels produced from biomass feedstocks for transportation applications include ethanol, methanol, and their ethers, ETBE and MTBE, as well as hydrogen, syngas, biodiesel, and jet fuels. Ethanol is produced from sugar (e.g., sugar cane), starch (e.g., corn), or cellulosic feedstocks (wood, herbaceous material, and MSW). The conversion process involves the biochemical conversion of the feedstock into its constituent glucose chains, which are then fermented to produce alcohol.

Methanol is produced from biomass by first gasifying the feedstock to form a syngas, a subsequent gas-shift reaction to adjust the chemical composition of the gas, cleaning, and finally the conversion to methanol in the presence of a catalyst. Hydrogen can also be manufactured via the gasification of biomass followed by gas-shift reactions and separation of the hydrogen component of the syngas.

Synthetic hydrocarbon fuels (gasoline, diesel, and jet fuel) can be created from biomass by pyrolizing the feedstock to form an intermediate biocrude liquid product and then catalytically converting the biocrude to a traditional hydrocarbon fuel formulation. Alternative methods convert the oils extracted from certain plant seeds to diesel fuel by isolating the hydrocarbon portion of the carbon chain.

Currently, ethanol is the only biofuel that has achieved noticeable market success, particularly in Brazil and, to an extent, in the U.S. The advanced technologies associated with producing ethanol from low-cost feedstocks (i.e., lignocellulosic biomass) may help enlarge the ethanol market after the turn of the century. Similarly, biomass-methanol production technologies might be commercially available in the first decade of the 21st century, particularly if gasification technologies continue to advance in the electric power sector. Finally, current research experience indicates that cost-effective hydrogen and synfuel production might also reach commercial levels sometime in the first decade of the 21st century.

- **Biomass Energy Systems: General Considerations**

In addition to the direct power and environmental benefits, biomass energy systems offer numerous other potential benefits, especially for developing countries. Some of these benefits include off-season employment of underutilized labor (tree planting and maintenance, use of tree co- and by-products); increased agricultural productivity through soil stabilization, reduced water runoff, and improved microclimates; and increased opportunities for industrial development through the availability of peak (wet season) and base load (dry season) power. In power deficit areas, the availability of power can encourage the development of small-scale industries and rural commerce, increase the productivity of agriculture through irrigation and post-harvest processing, make modern conveniences (e.g., lighting and potable water pumping) available to many more rural residents, and promote general well-being.

What separates the evaluation and implementation of biomass energy projects apart from other renewable and conventional energy systems is that resource or feedstock management is an integral part of a total energy, environmental, and economic system. Moreover, the complexity of the evaluation and possible implementation is further compounded by a high dependency of the biomass energy system on adaptation to local environmental, economic, social, and institutional considerations. Evaluation of this option must address a complex array of issues: competitiveness with other sources of energy supply, costs, emissions, financing, energy regulatory policies, food and fiber production, employment and impacts on agriculture, biodiversity and habitat, land ownership and tenure, flood and erosion control, pollution remediation, plantation sustainability, and logistics of resource supply and distribution.

9.2.4 Geothermal Energy

Geothermal energy systems tap the heat originating from the earth's molten interior and the decay of radioactive materials in the crust. The potential size of the resource is very large although conversion technologies for fully accessing the estimated 100 million quads of available worldwide resource are yet to be proven. Geothermal energy is currently being used in various locations around the world to produce electricity at costs competitive with conventional sources and provide energy directly for space heating, food and industrial processing, refrigeration, and aquaculture.

- **Geothermal Electric Technologies**

  Hydrothermal resources, consisting of water and/or steam trapped in fractured or porous rocks, are currently the only type of geothermal energy being accessed on a commercial scale. Electricity is produced via one of three major routes: dry steam, flash steam, and binary conversion. Dry steam systems use the geothermal steam directly to drive a turbine-generator while flash steam technologies first convert hot geothermal liquids to steam by quickly reducing its pressure. Finally, binary cycle systems are used for generating power from lower-temperature liquids by using the hot geothermal waters to vaporize a secondary working fluid which then drives a turbine-generator unit.

  The state of California in the U.S. currently receives 6% of its electricity from geothermal energy, and installed units exist in the Philippines, Mexico, Italy, Japan, New Zealand, and other countries. Advances in methods for locating, drilling, and extracting geothermal energy coupled with improvements in conversion technologies can help geothermal electricity expand its current market share further, effectively competing with fossil-powered sources in the baseload power market.

- **Geothermal Heating Technologies**

  In some regions of the world, low-temperature geothermal energy is being used directly for space heating, such as in several cities in Iceland where steam/hot water lines carry geothermal fluid through the district heating system. Another promising technology is the geothermal heat pump (GHP). GHPs operate like a conventional heat pump (a refrigerator, for example, is a one-way heat pump) and use the heat gradient between the earth's surface and groundwater or soil several hundred feet below the surface to power the pump. Because GHPs are reversible, they can provide space heating in the winter and space cooling in the summer as well as supplement domestic hot water needs year-round. More than 100,000 of these systems have been installed in the U.S. to date and sales continue to grow at significant rates.

9.2.5 Hydropower

Hydropower facilities exploit the kinetic energy in flowing or falling water to generate electricity. Conventional hydropower facilities use water from a river, stream, canal, or reservoir to continually produce electrical energy, and water releases from single-purpose reservoirs (i.e., dedicated to power production) can be quickly adjusted to match electricity loads. Multipurpose reservoirs are not capable of following load as closely; however, they can be simultaneously used for irrigation, flood control, navigation, recreation, and water supply.

Pumped storage plants operate similarly, but instead of tapping free-flowing water, the facility uses recycled water. During off-peak hours, water is pumped to an upper reservoir using low-cost resources, where it can be re-used to provide peak power on demand. Pumped hydropower facilities are net energy consumers (typically 1.25-1.40 kWh is required to pump the water to the upper reservoir for each kWh generated); however, they provide significant economic and operational benefits to utilities because of their ability to meet transient peak power demands.

Finally, mini-hydro facilities (30 MW or less in size) offer opportunities for distributed or remote power generation with minor environmental impacts, low operating costs, and high reliability. Installation of these systems is usually quite rapid and can use local labor.

Hydropower technology is currently mature and widely available. Almost 15% of the world's electrical energy comes from hydroelectric facilities operating in over 80 countries (Moreira and Poole 1993). Only a fraction of the available resource has been exploited to date, in large part because of siting constraints, environmental pressures against large-scale systems, and competition with other interests for water resource use. The increased development and use of micro-hydro technologies may permit additional resources to be accessed without encountering the barriers that have traditionally constrained conventional hydropower development. In Nepal, for example, the Agricultural Development Bank has financed the purchase of more than 650 small hydro systems by farmers, who use income from milling operations and electricity sales to pay back the loans.

9.2.6 Ocean Energy

Numerous systems have been conceived of for capturing the energy of the ocean's waves, tides, and temperature gradients. Of these, ocean thermal energy conversion (OTEC) and tidal power systems have received the greatest attention to date. OTEC uses the temperature difference between the surface of the ocean and depths of up to 1000 meters to generate electricity in either a closed-loop (using a secondary working fluid) or open-loop cycle (using seawater as the working fluid).

Tidal power uses the same principle employed in hydroelectric power generation to extract energy from a difference in hydrostatic head created by the rising and falling tides. A minimum difference of 5 meters between low and high tides is often cited as the limit required to effectively produce electricity, although developments in micro-hydro technology may permit additional resources with lower ranges to be accessed.

OTEC and tidal power are still primarily in the development stage, and technical, cost, and siting constraints continue to limit the progress of these systems.

9.2.7 Screening Mitigation Options

The first step in screening renewable energy options for further analysis is to conduct a preliminary resource assessment. For many countries, such an assessment may have already been conducted with sufficient accuracy to allow a screening out of those options that are unlikely to be of interest because adequate resources are lacking, or are too small to make a significant national contribution. Alternately, the study team can conduct its own preliminary assessment, drawing on appropriate experts in the country. For assessment of biomass resources, the energy sector analyst should work together with the forestry and agricultural analysts.

For those technologies for which exploitable resources appear to be available, a screening should be conducted using the criteria listed in Table 2-1 in Chapter 2. It is particularly important to consider the potential for indirect economic benefits such as local employment creation and foreign exchange savings.

9.3 RESOURCE ASSESSMENT[1]

Use of renewable energy requires careful matching of conversion technology to resource availability and to energy demand. Although it is usually obvious to the inhabitants of an area that one form of renewable energy is more plentiful than another, it is important to conduct an inventory of renewable energy resources before selecting technologies that may be deployed to convert these resources to usable energy forms. Some countries may have progressed beyond what is discussed below on resource assessment, while others may not have.

Resource assessment requires determination of the quantity and quality of resources found in a given region at a given time. It may involve exploration and monitoring to determine whether a particular site is suitable for a particular renewable energy technology.

An inherent difference between renewable and conventional energy resources is that renewable sources are determined by their "flow" whereas conventional sources are "fixed" in stock. Renewable energy resources vary by time of day, and some of them, such as solar radiation and wind, vary by night and day. Most renewable energy sources vary seasonally, and all of them vary from year to year. Sunlight and wind are variable in place, also. The rate at which renewable energy resources can be expended is generally fixed by the flow and not, as with conventional resources, by demand. These inherent resource characteristics make it essential to carefully evaluate energy needs when considering renewable energy systems. Some renewable energy resources, such as biomass resources, have competing uses that must be considered. Furthermore, crops can be grown for energy as well as agricultural uses, and in cases such as bagasse from sugar cane, there are mutual benefits in using a biomass resource for both agricultural and energy applications.

Data availability for renewable energy resources varies widely. Furthermore, most of the data available have been collected primarily for other purposes such as agriculture, airline operations, flood control, timber harvests, waste management, and the protection of life and property (weather forecasts). Such data are valuable for those purposes, but they do not always provide the information required to make accurate assessments of renewable energy resources. The collection of data specifically for assessing renewable energy resources is quite limited. In addition, data that are available for renewable energy resources are generally lacking in accuracy and spatial (geographic) and temporal (time) resolution. The poor spatial resolution (lack of data at most locations) is the direct result of the small number of locations at which resource measurements have been made. The lack of temporal resolution means that it is not possible to accurately estimate the flow characteristics of renewable resources on time scales capable of resolving hourly, daily, monthly, and annual variations.

9.3.1 Solar Radiation

Most measures of solar radiation resources are given in terms of the total energy received (in Btu, kWh, or megajoules) each hour and/or day at the measurement location. Monthly and annual averages are computed from the hourly or daily data, and these are used to prepare tabular data sets and maps of average daily energy. These data tables and maps provide a first-level indication of resource availability.

---

[1] The discussion on resource assessment is taken from Maxwell and Renne (1994).

More detailed assessment requires resolving the temporal characteristics of solar radiation—changes that take place over time scales ranging from minutes to decades. Experience has shown that short-term changes in the flow of solar energy requires measurement intervals of 5 minutes or less. Monitoring long-term changes associated with climate change, volcanic eruptions, and atmospheric pollution requires a continuous measurement program. Typically, solar radiation data have been collected at intervals from one minute to one day, with the most common interval being one hour. Hourly data are adequate for most applications. The short-term (e.g., five-minute) variability of solar radiation at a given site can usually be determined from one to three years of data.

For initial screening of solar radiation resources and to produce solar resource maps at the national level, monthly average daily totals of the global horizontal radiation are available on Internet from a database at NASA-Langley at a resolution of 280 km by 280 km grids. These data are available for all locations. These data can be processed to obtain finer spatial (to 30 km grids) and temporal resolution, but such work will require specific effort by experts. These data should be validated by ground-based observations before they are applied to evaluate the suitability of solar energy technologies or projects to meet specific end uses or needs. International solar radiation data are available from the World Radiation Data Center (WRDC) in St. Petersburg, Russia, and from the World Radiation Monitoring Center in Zurich. Data from the WRDC will be available in the near future from The National Renewable Energy Laboratory (NREL) on Internet.

9.3.2 Wind Energy

Because the power of the wind is proportional to the cube of the wind speed, a small change in wind speed can represent a large change in wind power at any given time. Wind power density is calculated by summing the average wind power of several ranges of wind speed over a specified period of time (month, season, or year). The power for each range is multiplied by the frequency of occurrence of wind speeds within that range, and the results are summed over all ranges to obtain the wind power density. Maps based on such wind speed ranges to define wind power classes are available but at such large scales that even initial screening at the national level for most countries will be difficult. Wind atlases for many parts of the world are available but vary in quality and consistency. Grubb and Meyer (1993) survey data availability at the national and continental levels. Since wind speed is highly variable and is affected by surface roughness and topography, regional or local data are desirable for screening purposes.

Similar to solar radiation, wind speed has not generally been measured for the purpose of assessing wind energy resources. For these reasons and others, most of the wind data available today are not as reliable as one would like for estimating wind energy resources. Wind resources, including the turbulence characteristics of the resource, can vary significantly over small distances, particularly in very mountainous terrain or in coastal or lake-shore regions. When no wind data are available, topographic and vegetative indicators, or numerical models, might be used to estimate wind resources.

9.3.3 Hydropower

The hydropower resources for a given river basin or catchment area can be estimated from three sources of information: hydrographic records, which define the rate of flow of water at a certain point within the river basin, measurements or estimates of precipitation falling on the catchment area, and the topographic characteristics of the catchment basin. Recording rain gauges are standard for virtually all weather service stations around the world.

The prospects of developing any site also depend on numerous engineering, environmental, and economic considerations which have not been fully assessed and which can change over time. In those areas of the world with rapid population growth or areas that are prone to periodic severe droughts, there can be intense competition for the available freshwater supply. As with other renewable resources, the cost of energy derived from a hydropower project can vary considerably, even within regions where the resource is generally abundant.

9.3.4 Geothermal

Geothermal energy resources fall into four categories: hydrothermal (hot water and steam), hot dry rock, geopressured hot water with dissolved CO_4, and magma. Only hydrothermal resources have been commercially developed to date. Palmerini (1993) provides an overview of global geothermal resources and extraction technologies.

The main characteristics of a geothermal reservoir that determine the usefulness of its energy are its volume, the water state (liquid and/or steam), temperature, pressure, depth, water chemistry, and recharge rate. High-temperature steam sources are of greatest value for generating electricity. Lower temperature liquid sources ($T< 90°C$) are typically used for heating buildings and for process heat. The depth and pressure, particularly for liquid resources, will affect the cost of getting the resource to the surface. The chemistry of the liquid or steam will affect construction and operational costs, according to the measures required to remove harmful minerals or mitigate their effects. The rate at which the hot water or steam is replenished (naturally and by injecting spent fluids) determines the quantity of energy that the source is capable of supplying on a continuous basis.

9.3.5 Biomass

Biomass is by far the most complex of the renewable energy resources and presents the most difficult assessment problems. This complexity is due primarily to the multiple sources of biomass—including energy crops (e.g., trees, corn, sugarcane, microalgae); standing crops (mostly forests) grown for other purposes; litter and dead or non-commercial trees (standing waste) in forests; and forest and agricultural wastes (field and mill or factory)—and to the many energy conversion processes that are possible, including direct combustion, gasification, liquefaction, and biochemical processing. The first step (after determining end-use needs) is to inventory sources of biomass and their availability. Hall, *et al.* (1993) reviews biomass resources at the national and continental levels.

Once the sources and availability are identified, harvesting rates, transportation, storage, and conversion issues must be considered. For example, microalgae might be harvested on a daily basis, agricultural crops and agricultural waste from one to three times per year, trees grown for energy use every three to five years. It may be economical to harvest litter and standing waste only in conjunction with a timber harvest, which may occur as infrequently as once every 10 to 50 years. Given the variety of biomass resources, several resource assessment methods will be needed. *Assessment of biomass resources for the energy sector should be coordinated with the non-energy part of the mitigation assessment.*

Biomass resources are also difficult to assess because they consist of both flow and fixed stock components. Standing crop and litter in a mature forest represent a fixed stock waiting to be harvested. Annual growth in young forests, agricultural crops, and harvesting and production wastes all represent flow resources that are subject to change seasonally and from year to year. With the exception of crops grown for energy use, there is little consensus regarding the proportion of most biomass resources that is available for energy use. Competing uses of biomass for food, clothing, and shelter also complicates resource assessment. Competition for many biomass waste products is also important to consider. Wastes from both forest and agricultural crops are essential to stabilize and replenish soils. Straw and wood chips are used for bedding material for animals, and some waste products, such as bean pods and sugar beet pulp, are used as animal feed. In general, the amount of waste that could be diverted to sustainable energy production is not well known.

9.3.6 Municipal Solid Waste

Municipal solid waste (MSW) can be defined as the post-consumer solid waste generated by residential, commercial, and institutional (schools, hospitals, offices, etc.) sources (OTA 1989). Industrial sources are not included because industrial wastes are usually discarded separately from municipal wastes. Some of these solid wastes are combustible and are being used in some areas as a renewable source of energy. The feasibility of using MSW for energy production in a given area depends on a number of factors, including population, availability of landfills, the general economy of the area, and the demand for thermal or electrical energy.

An estimate of the national resource base requires an estimate of the materials flow of paper and paperboard, plastics, yard trimmings, rubber and leather, textiles, wood, food wastes, and other categories of MSW that could be used as an energy resource. Estimates of MSW resources for individual municipalities requires measurements of the MSW at the dump site or the point of generation. At the dump, the total weight could be determined by weighing the trash collection trucks before and after dumping. Measuring the composition of the MSW is more costly and time-consuming. A certain percentage of the trash trucks must be randomly selected and their contents dumped and sorted into categories such as paper and paperboard, plastics, wood, metal, etc. Each category must then be weighed separately. Similar to any other measurement process, the accuracy and reliability of MSW sampling is dependent on the design of the sampling program (e.g., number of samples, distribution in time, and length of sampling period) and its execution. Currently, there is no standard method for collecting such data; in fact, there is no general agreement on the categories to be used to define MSW (Kahn and Sable 1988).

9.4 TECHNOLOGY CHARACTERIZATION

After renewable energy resources are assessed to the degree possible with the available data, the next step is to characterize technologies that can convert the resources to usable forms. Renewable energy technologies are as varied as the resources they convert to electricity, heat, and fuels, and the information required to evaluate them differs somewhat among technologies. In general, however, the types of data listed in Table 8-11 for conventional energy supply technologies (engineering performance, economic parameters, environmental impacts) also apply for renewable energy technologies.

The National Renewable Energy Laboratory (NREL) in the U.S. has prepared technical and economic descriptions (technology characterizations) of selected renewable energy technologies for the IPCC that include performance and cost data. These technology characterizations include those renewable energy technologies considered by the Energy and Industry Subgroup of the IPCC to be commercially available in the near-term (<10 yrs) and suited for limiting the levels or growth rates of greenhouse gas emissions in developing and transitional economies. The technologies characterized and their applications are:

- flat-plate photovoltaics (electricity)
- solar thermal (heat, electricity)
- salt-gradient ponds (electricity, space and process heat, desalinization)

- wind turbines (electricity, water pumping)
- small-scale hydro (electricity)
- geothermal (electricity)
- MSW (electricity, heat)
- biomass: direct-fire steam turbine, gasification-gas turbine cycles (electricity)
- pelletized wood (space and process heating).

Additional cost information on photovoltaics, solar thermal electric, and biomass energy (electricity and liquid fuels) is available in Ahmed (1994).

An important set of technologies not included in the IPCC work so far is that to produce liquid fuels from biomass. Technology characterizations of these and other energy systems listed in Table 9-1 are available from NREL.

9.5 CONSTRUCTING BASELINE AND MITIGATION SCENARIOS

After potentially applicable renewable energy technologies are identified, a method to select the most suitable technologies is needed. There are two general ways to make this selection. One way is to use optimization or other models that can integrate renewable energy technologies into the overall energy supply system once assumptions on costs and performance are made.[2] The second way is to use a decision analysis tool that explicitly accounts for subjective and non-quantifiable factors. An example of such a tool is the Analytical Hierarchy Process (AHP), which has been applied for evaluation of renewable energy technologies in Mexico (Corbus *et al. 1993).*

Specific options can be further analyzed by identifying market opportunities and exploring market penetration options. In doing so, attention should be paid to some of the unique characteristics of renewable supply technologies. Renewables offer important benefits in terms of reduced environmental impact beyond climate change mitigation, energy independence and diversity, and economic and social development. These benefits are often not readily modeled but are essential elements of a comprehensive assessment.

A particularly promising area for renewables deployment is distributed or off-grid power generation. Accounting for these opportunities requires a way of evaluating the trade-off between grid extension and off-grid remote power generation. In many instances, a utility line extension can be more costly than deploying a renewable technology as a stand-alone system.

For on-grid applications, a key issue for renewable energy is integrating the technologies into the existing energy network. Many of the renewable technologies operate intermittently, which is to say that they only produce power part of the time (when the sun is out or the wind is blowing). Some of the outages can be predicted (e.g., the time that the sun will set is known), while others cannot (e.g., a large bank of clouds passing overhead). As a result, electric utilities consider most solar and wind technologies as non-dispatchable since they cannot be turned on whenever they are required. Considerable research and experience has been devoted to assessing how utilities can deal with this uncertainty. One method has been to increase the knowledge of the resource through resource assessment to reduce the uncertainty

---

[2] Many models assess the availability of resources (e.g., oil, gas, coal) through pricing. As the resource becomes more scarce, the fuel price of a technology increases and the deployment of the technology is therefore constrained. For many renewables, fuel price does not exist (i.e., solar or wind technologies) and there is no way for the model to automatically limit deployment based on resource availability. In these instances, a "cap" may need to be applied to the model.

about when the resource is available (i.e., "predict" when the wind will come up). Other techniques include the use of backup and storage systems, promoting system diversity, and load management.

9.6 MITIGATION POLICIES

Policies to promote adoption of renewable energy options and examples of their application are briefly described in Table 9-2. These are loosely aligned with the policy options listed in Chapter 8 on conventional energy supply since many of the same types of policies will serve to enhance renewable energy supply. Table 9-2 only lists policies and examples that might directly promote the use of renewable energy; however, most of the options discussed in Chapter 8 can have indirect effects on the use of renewables. For example, taxes or restrictions on the use of carbon-intensive fuels (e.g., coal) will increase opportunities for renewable energy. Finally, integrated resource planning offers opportunities for incorporating renewable energy technologies into utility planning as part of a formal process.

In addition to the policies in Table 9-2, technological transition strategies should be part of an implementation plan for renewable energy. Hybrid systems such as diesel generator/wind, hydro pumped storage/wind, and co-firing of renewable and conventional fuels (e.g., biomass and coal) are ways of integrating renewable and conventional technology options.

Table 9-2. Policies to Promote Use of Renewable Energy

OPTION	DESCRIPTION	EXAMPLES
Emissions Controls	Limits on emissions of GHG or other pollutants - caps by technology - caps by region/sector	- Emissions cap on new power plants, transportation vehicles, etc. - Emissions cap from regional utilities, urban zones (e.g., "clean cites"), or segments of the industrial sector
Input Controls	Require uses of renewable energy feedstocks	- Set-asides for renewable utility technologies in future capacity expansion plans - Payments to operators using renewable energy for industrial processes
Price Controls	Price subsidies for renewable energy fuels and technologies	- Federal/local subsidies on cost of electricity from renewable utility technologies - Payments to operators using renewable energy for industrial processes
Rate-of-Return Regulation	Requirements on the rate-of-return calculation for energy supply companies	- Permit owners/operators of renewable electric technologies to achieve higher rates of return
Permitting	Conditions placed on energy supply operations	- Preferential siting of renewable energy supply facilities; reduce siting constraints
Tradeable Emission Permits	Require the acquisition of GHG emission rights by trading with other sources	- Renewable energy supply technology used to offset emissions from non-renewable sources
Taxes	Tax credits/levies for GHG control - on products - on firms - on land and facilities	- Tax credits for renewable energy use (fuels, electricity, etc.) - Taxes on GHG-emitting technologies
Tariffs	Reduced tariffs for importation of non-GHG emitting technologies	- Reduced tariffs on foreign renewable energy supply technologies (e.g., wind energy systems) or demand tech-nologies (e.g., alcohol-fueled vehicles)
Research, Development, and Deployment	Government funding of RD&D in the area of renewable energy	- Government-sponsored research, including demonstration activities
Government Purchase	Government purchase of renewable energy technologies	- Purchase of alternative-fueled vehicles for government fleet use - Cost-sharing with industry and in purchase of industrial renewable energy technologies
Information Dissemination	Promote renewable energy technologies with the general public, specific industry groups, and energy planners	- Creation of one-stop shopping information "clearinghouses" for renewables - Sponsorship of technical conferences with industry
Cooperatives	Promote cooperatives that aggregate demand for renewable energy technologies	- Rural electric cooperatives (with focus on village power) - Industry-sponsored cooperatives (e.g., pulp and paper industry)

REFERENCES

Ahmed, K., 1994: Renewable Energy Technologies: A Review of the Status and Costs of Selected Technologies, World Bank Technical Paper No. 240, Washington, DC: The World Bank.

Capentieri, A.E., E.D. Larson, and J. Woods. 1993. "Future Biomass-Based Electricity Supply in Northeast Brazil," *Biomass and Bioenergy*, 4:149-174.

Grubb, M.J. and N.I. Meyer. 1993: "Wind Energy: Resources, Systems and Regional Strategies", in Johansson, *et al.*, 1993.

Hall, D.O., *et al.* 1993: "Biomass for Energy: Supply Prospects", in Johansson, *et al.*, 1993.

Kahn, Z.K. and E. Sable. 1988: "Planning a Program to Determine Physical and Chemical Characteristics of Municipal Solid Waste," Resource Recovery/Cogeneration World, 1:15-18.

Johansson, T.B., *et al.* 1993: Renewable Energy: Sources for Fuels and Electricity, Washington, DC: Island Press.

Maxwell, E.L. and D.S. Renne. 1994: Measures of Renewable Energy, National Renewable Energy Laboratory (NREL/MP-463-6254), Jan. 1994.

Moreira, J.R. and A.D. Poole. 1993: "Hydropower and Its Constraints", in Johansson, *et al.* 1993.

OTA. 1989: Facing America's Trash – What Next for Municipal Solid Waste? OTA-O-424, U.S. Congress, Office of Technology Assessment, Washington, DC: U.S. Government Printing Office.

Palmerini, C.G. 1993: "Geothermal Energy", in Johansson, *et al.* 1993.

Perlack, R.D., Ranney, J.W. and M. Russell. 1991. *Biomass Energy Development in Yunnan Province, China*, Oak Ridge National Laboratory, ORNL/TM-11791, Oak Ridge Tennessee, June 1991.

Russell, M., Jantzen, D. and Z. Shen. 1992. *Electricity From Biomass: Two Potential Chinese Projects*, Energy, Environment, and Resources Center, University of Tennessee, Knoxville, 1992.

CHAPTER 10
INTRODUCTION AND A LAND-USE FRAMEWORK FOR THE NON-ENERGY SECTORS

10.1 INTRODUCTION

Chapters 10 through 14 of this book cover the three main land-based sectors—forestry, agriculture, and rangelands and grasslands—and waste management. The land-based sectors are responsible for emissions of CO_2 associated with land-use changes, and they can also sequester carbon through photosynthetic processes. Together with the waste management sector, these sectors are also responsible for most of the anthropogenic emissions of the GHG methane, non-methane hydrocarbons, NO_x, and oxides of N_2. These sectors therefore provide opportunities for mitigating climate change through reduction of GHG emissions and/or through increased absorption and storage of carbon in perennial vegetation, detritus, soils, and in long-term biomass products.

In the forestry sector, the main mitigation options involve: 1) reducing emissions by maintaining existing carbon sinks through conservation and protection, efficiency improvements, and fossil fuel substitution, and 2) sequestering carbon by expanding vegetation cover and increasing carbon storage in soils and in long-term products.

In the agriculture sector, one option involves the adoption of agricultural practices that increase the amount of crop residue on or in the surface soil layer; this reduces maximum temperatures and evaporation and increases organic matter in the soil, thereby trapping carbon in the soil. Other agricultural options include reduction of GHG emissions from animal husbandry, rice production, and fertilizer application.

In rangelands and grasslands, most options are based on enhancement of carbon sinks through improved range management, wild fire management, animal husbandry, and biomass replenishment.

The main options in the waste management sector involve reducing methane emissions from landfills and waste-water treatment.

10.2 ESTABLISHING LAND-USE PATTERNS FOR MITIGATION ANALYSIS

The availability of land and its products shape the mitigation options in the non-energy sectors. Forests, range and grasslands, and agricultural lands together comprise most of a country's vegetative land area. Changes in the land-use pattern in one sector directly influence the magnitude of land available for the other sectors. Land conversion from forests to agriculture or to pasture lands is a common feature in many developing countries. Where afforestation programs are active, wastelands are being reforested to yield industrial wood products. Elsewhere, the construction of dams and urban sprawl are taking over what were once fertile agricultural lands. Ultimately, the amount of land area available in a country is fixed, and the bounty of the land has to be shared among its various users. Exactly how the lands are shared will be determined by land-use demand and supply factors, as moderated by land-use policy and law. Moreover, the options in each sector are influenced by cross-sectoral issues rooted in the country's land-use policy and law.

The regulations governing the distribution of land in one sector are often connected to those for the other sectors. Because of these interactions, it is important that the availability of land and its distribution for alternative activities be evaluated in a comprehensive manner with the participation of specialists on forestry, agriculture, and range and grasslands.

The existing models that incorporate land-use change are either small-scale, site-specific models for which applicability to the whole economy is difficult[1], or broad-based regional bio/physiochemical process models, the output of which can only be used as an intermediate input in simulating land-use redistribution among sectors.

For the purposes of mitigation analysis, the simple framework presented below can be used for tracking land use change over several decades. The framework requires exogenous inputs to relate current and future land use activity and product demand and supply. The most common factors that affect land use and its products include:

- Demographic variables such as human population and its growth rate, rural/urban distribution, and dependence on land resources

- Economic factors such as income level, technological development, dependence on export of land-based products, and rates of economic growth

- Type and intensity of land use, such as shifting versus permanent agriculture, or clear-cutting versus selective harvesting

- Biophysical factors such as soil productivity, topography, and climate.

Using this framework for a mitigation evaluation consists of two steps: 1) evaluating current and future land availability for the baseline and mitigation scenarios, and 2) determining the demand for land-use products and reconciling the availability of land to meet product demand.

Step 1: Evaluate land availability for the baseline and mitigation scenarios.

Step 1.1: It is useful to begin by describing the main policy and land tenure elements which determine the existing distribution among the three sectors, and to identify the likely trends in land-use change. The latter may include increased acquisition of forest land for agriculture or pasture, abandonment of agricultural land for pasture or to secondary forests, or loss of primary forest lands to severely or permanently degraded lands. The analyst should define a *baseline* scenario using these elements as a basis.

Table 10-1 shows the land area changes for a hypothetical country. In this example, range and grasslands are considered a subset of forest land. (This example is also used to illustrate the evaluation of the forestry sector mitigation options in Chapter 11.)

Current land area statistics are organized as shown in the table. Government plans and other ongoing trends of afforestation, land conversion, etc. may be used to determine a baseline mosaic of future land use patterns. Plans for land area changes are usually not available beyond a 5-10 year time horizon. These may be extrapolated to estimate each type of land area up to 2030.

Step 1.2: A land-use pattern for a *mitigation* scenario may be constructed using the above baseline trends as a starting point. The example assumes future changes in land use in order to illustrate two types of *forestry* mitigation options. The first option is to *protect* 12,000 ha of current national park land, which is being depleted to 1000 ha in the baseline scenario by 2030. The second option is to *reforest* 40,000 ha of wastelands into forest land.

---

[1] One such model is LUCS, which has a project-level focus, with a capability for cost-benefit analysis (Faeth *et al.*, 1994).

TABLE 10-1
Establishing Land-Use Patterns for Mitigation Analysis

	1990	1991	1992	2010	2030
>>>> STEP 1.1: LAND USE PATTERN, BASELINE SCENARIO					
Thousands of Hectares					
>>> Forest Land					
>> >40% Crown Cover	100	100	100	100	100
>> 10-40% Crown Cover (	50	50	50	50	50
>> <10% Crown Cover					
> Rangelands	25	25	25	25	25
> Grasslands	20	20	20	20	20
> Wastelands	40	40	40	40	40
>> Other					
>>> Protected Land					
>> Wildlife Sanctuaries	10	10	10	10	10
>> National Parks	12	11.7	11.5	6.5	1
>> Other					
>>> Crop Land					
>> Perrenial	1,000	1,000	1,000	1,000	1,000
>> Annual	1,500	1,500	1,501	1,506	1,511
>> Shifting	200	200	200	200	200
>> Current Fallow	1,200	1,200	1,200	1,200	1,200
>> Other					
>>> Other					
>> Urban	1	1	1	1	1
>> Dams and Roads	300	300	300	300	300
>> Mines	50	50	50	50	50
>>> Not Classified above	20	20	20	20	20
>>>> TOTAL (million ha.)	4,528	4,528	4,528	4,528	4,528
>>>> STEP 1.2: LAND USE PATTERN, MITIGATION SCENARIO					
>>> Forest Land					
>> >40% Crown Cover	100	101	102	120	140
>> 10-40% Crown Cover (	50	50	50	50	50
>> <10% Crown Cover					
> Rangelands	25	25	25	25	25
> Grasslands	20	20	20	20	20
> Wastelands	40	39	38	20	0
>> Other					
>>> Protected Land					
>> Wildlife Sanctuaries	10	10	10	10	10
>> National Parks	12	12	12	12	12
>> Other					
>>> Crop Land					
>> Perrenial	1,000	1,000	1,000	1,000	1,000
>> Annual	1,500	1,500	1,500	1,500	1,500
>> Shifting	200	200	200	200	200
>> Current Fallow	1,200	1,200	1,200	1,200	1,200
>> Other					
>>> Other					
>> Urban	1	1	1	1	1
>> Dams and Roads	300	300	300	300	300
>> Mines	50	50	50	50	50
>>> Not Classified above	20	20	20	20	20
>>>> TOTAL (million ha.)	4,528	4,528	4,528	4,528	4,528

Step 2: Determine or estimate the demand for products (fuelwood, timber, crops, etc.) of land use and reconcile the land availability with the demand.

Step 2.1: The 1990 forest products demand is shown in the example (Table 10-2). Each type of land area is a source of several products. If mitigation options are to be successful, it is imperative that the future demand for these products be satisfied either through more intensive use of land or through imports of products. Alternatively, countries may also wish to pursue a strategy to replace non-wood products with wood ones. In our example, we illustrate such a scheme whereby wood produced from reforested wastelands is used to replace non-wood industrial products. Countries exporting timber products would also want to ensure that revenue earned from exports does not diminish because of the implementation of options.

The example shows fuelwood, industrial wood, agricultural products, livestock, and electricity generation (hydro) as the products that might be derived from land. The analyst should include products appropriate to his/her country.

For 1990, the example shows the absolute demand for, and the supply of, each product and its intensity of land use. It is important to ensure that the land-use intensity does not exceed well- known norms for each type of product. These norms may be obtained in consultation with specialists in the country.

Step 2.2: The baseline demand for these products may be projected to 2010 through the use of econometric models, end-use projections, and other techniques. Alternatively, estimates of demand may be obtained from experts in the field. For the baseline scenario, the example shows the absolute product demand and supply from various types of land use and the intensity of land use in 2010 for each product (Table 10-3).

Step 2.3: Table 10-4 shows the effect of the two forestry mitigation options on product supply for 2010. The non-wood industrial products in the baseline scenario are replaced by production from reforested wasteland in the mitigation scenario. The protection of national parks results in a loss of 40,000 tonnes of agricultural products which are imported in the mitigation scenario. The evaluation of these two options, reforestation and protection, is described in Chapter 11.

Similar analyses should be conducted for 2030 for both scenarios. If the demand for the products cannot be met, then either the mitigation options will need to be modified or the product may need to be imported.

Use of the framework described above requires interaction among experts working in the forest, agriculture, range and grasslands, and energy sectors. Each sectoral specialist should construct his/her own baseline and mitigation scenarios and reconcile the projected overlapping use of land with experts from the other sectors. This process will yield a consistent set of baseline and mitigation land-use scenarios.

TABLE 10-2. Supply of, and Demand for, Land-Use Products, Country X, 1990

	>>>> STEP 2.1: 1990 PRODUCT SUPPLY AND DEMAND					>>>> 1990 INTENSITY OF LAND USE (/ha)				
	Fuel Wood '000 t	Indus. Wood '000 t	Agric. Product '000 t	Live-stock '000	Elec-tricity GWh	Fuel Wood t/ha	Indus. Wood t/ha	Agric. Product t/ha	Live-stock #/ha	Elec-tricity MWh/ha
>>> Forest Land										
>> >40% Crown Cover		10,000				0.00	100.00	0.00	0.00	0.00
>> 10-40% Crown Cover (Woodl	50	1,000				1.00	20.00	0.00	0.00	0.00
>> <10% Crown Cover										
> Rangelands				5		0.00	0.00	0.00	0.20	0.00
> Grasslands				10		0.00	0.00	0.00	0.50	0.00
> Wastelands				10		0.00	0.00	0.00	0.25	0.00
>> Other										
>>> Protected Land										
>> Wildlife Sanctuaries						0.00	0.00	0.00	0.00	0.00
>> National Parks	20		40			1.67	0.00	3.49	0.00	0.00
>> Other										
>>> Crop Land										
>> Perrenial	1,000		1,200	260		1.00	0.00	1.20	0.26	0.00
>> Annual	500		1,700			0.33	0.00	1.13	0.00	0.00
>> Shifting	15		300	30		0.08	0.00	1.50	0.15	0.00
>> Current Fallow						0.00	0.00	0.00	0.00	0.00
>> Other										
>>> Other										
>> Urban						0.00	0.00	0.00	0.00	0.00
>> Dams and Roads					45	0.00	0.00	0.00	0.00	0.15
>> Mines						0.00	0.00	0.00	0.00	0.00
>>> Not Classified above						0.00	0.00	0.00	0.00	0.00
>> Additional Categories										
>> Additional Categories										
>> Imports/ (Exports)										
>>> TOTAL SUPPLY OF PRODUCTS	1,585	11,000	3,240	315	45	0.35	2.43	0.72	0.07	0.01
>>> TOTAL DEMAND FOR PRODUCTS	1,585	11,000	3,240	315	45					

TABLE 10-3. Supply of, and Demand for, Land-Use Products, Country X, 2010

>>>> STEP 2.2: 2010 BASELINE SCENARIO: PRODUCT SUPLY AND DEMAND						>>>> 2010 BASELINE SCENARIO: LAND USE INTENSITY				
	Fuel Wood '000 t	Indus. Wood '000 t	Agric. Product '000 t	Live-stock '000	Elec-tricity GWh	Fuel Wood t/ha	Indus. Wood t/ha	Agric. Product t/ha	Live-stock #/ha	Elec-tricity MWh/ha
>>> Forest Land										
>> >40% Crown Cover		10,000				0.00	100.00	0.00	0.00	0.00
>> 10-40% Crown Cover (Woodl	400	1,000				8.00	20.00	0.00	0.00	0.00
>> <10% Crown Cover										
> Rangelands	3			5		0.12	0.00	0.00	0.20	0.00
> Grasslands				10		0.00	0.00	0.00	0.50	0.00
> Wastelands				10		0.00	0.00	0.00	0.25	0.00
>> Other	0									
>>> Protected Land										
>> Wildlife Sanctuaries						0.00	0.00	0.00	0.00	0.00
>> National Parks	20		40			1.67	0.00	3.49	0.00	0.00
>> Other										
>>> Crop Land										
>> Perrenial	1,000		2,000	288		1.00	0.00	2.00	0.29	0.00
>> Annual	1,450		4,871	156		0.97	0.00	3.25	0.10	0.00
>> Shifting	15		300	30		0.08	0.00	1.50	0.15	0.00
>> Current Fallow						0.00	0.00	0.00	0.00	0.00
>> Other										
>>> Other										
>> Urban						0.00	0.00	0.00	0.00	0.00
>> Dams and Roads					100	0.00	0.00	0.00	0.00	0.33
>> Mines						0.00	0.00	0.00	0.00	0.00
>>> Not Classified above						0.00	0.00	0.00	0.00	0.00
>> Additional Categories										
>> Additional Categories										
>> Imports/ (Exports)		0	0							
>>> TOTAL SUPPLY OF PRODUCTS	2,888	11,000	7,211	499	100	82.2%	0.0%	122.6%	58.4%	122.2%
>>> TOTAL DEMAND FOR PRODUCTS	2,888	11,000	7,211	499	100					

TABLE 10-4. Supply of, and Demand for, Land-Use Products, Country X, 2010

>>>> STEP 2.3: 2010 MITIGATION SCENARIO: PRODUCT SUPPLY AND DEMAND						>>>> 2010 MITIGATION SCENARIO: LAND USE INTENSITY				
	Fuel Wood '000 t	Indus. Wood '000 t	Agric. Product '000 t	Live- stock '000	Elec- tricity GWh	Fuel Wood t/ha	Indus. Wood t/ha	Agric. Product t/ha	Live- stock #/ha	Elec- tricity MWh/ha
>>> Forest Land										
>> >40% Crown Cover		10,000				0.00	100.00	0.00	0.00	0.00
>> 10-40% Crown Cover (Woodl	400	1,000				8.00	20.00	0.00	0.00	0.00
>> <10% Crown Cover										
> Rangelands	3			5		0.12	0.00	0.00	0.20	0.00
> Grasslands				10		0.00	0.00	0.00	0.50	0.00
> Wastelands		2,500		10		0.00	125.00	0.00	0.25	0.00
>> Other	0									
>>> Protected Land										
>> Wildlife Sanctuaries						0.00	0.00	0.00	0.00	0.00
>> National Parks	20					1.67	0.00	0.00	0.00	0.00
>> Other										
>>> Crop Land										
>> Perrenial	1,000		2,000	288		1.00	0.00	2.00	0.29	0.00
>> Annual	1,450		4,871	156		0.97	0.00	3.25	87.84	0.00
>> Shifting	15		300	30		0.08	0.00	1.50	0.15	0.00
>> Current Fallow						0.00	0.00	0.00	0.00	0.00
>> Other										
>>> Other										
>> Urban						0.00	0.00	0.00	0.00	0.00
>> Dams and Roads					100	0.00	0.00	0.00	0.00	0.33
>> Mines						0.00	0.00	0.00	0.00	0.00
>>> Not Classified above						0.00	0.00	0.00	0.00	0.00
>> Additional Categories										
>> Additional Categories										
>> Imports/ (Exports)		0	40							
>>> TOTAL SUPPLY OF PRODUCTS	2,888	13,500	7,211	499	100	82.2%	22.7%	121.3%	58.4%	122.2%
>>> TOTAL DEMAND FOR PRODUCTS	2,888	13,500	7,211	499	100					

REFERENCE

Faeth, P., Cort, C. and Livernash, R. 1994. Evaluating the Carbon Sequestration Benefits of Forestry Projects in Developing Countries. Washington, D.C.: World Resources Institute and U.S. Environmental Protection Agency.

CHAPTER 11
FORESTRY SECTOR

11.1 INTRODUCTION

Forestry mitigation options refer to those measures and policies that can lead to a reduction in the emission of greenhouse gases from forestry and/or increase carbon sequestration in forests, long-term wood products, and other tree vegetation. In most land use changes involving decomposition and oxidation, GHGs will be emitted into the atmosphere. These gases include CO_2, CO, methane, N_2O, NO_x and other NMHC. Although CO_2 accounts for the bulk of these gases emitted from the forest sector, it can also be re-absorbed by vegetation, soils and water bodies. The other GHGs, while emitted in trace amounts, accumulate in the atmosphere for their entire residence period without a possibility for re-absorption. The emissions of these trace gases can be estimated using standard ratios to the amounts of CO_2 released.

11.2 MITIGATION OPTIONS

Mitigation options for the forestry sector may be classified into two basic types. One type involves expanding the pool of carbon in soils, vegetation, and wood products. Expansion withdraws carbon from the atmosphere and sequesters it. The second type involves maintaining the existing pools of carbon in soils, vegetation, and the proportion of forest products currently in use. Maintaining existing stands, whether achieved through reduced deforestation, forest protection, or more efficient conversion and use of forest products, keeps the avoided GHG emissions from entering into the atmosphere for the duration of the pool maintenance. Although expansion and maintenance of carbon pools in standing trees, forest soils, and forest products may be very effective mitigation options, they may be difficult to implement since the alternative use of the land upon which the carbon is stored is often more valuable to local inhabitants than the trees are.

Another way to reduce carbon emissions is to use wood obtained from renewable sources like forest plantations as a substitute for non-renewable emission sources, particularly fossil fuels (Hall *et al.*, 1991). This substitution will delay the release of carbon from the fossil fuel for as long as one continues to use wood from a renewable source in lieu of the fossil fuel. Similarly, wood derived from renewable sources, if used as a substitute for woodfuel derived from depletable natural forests, will also delay or stop carbon release from the non-renewable source. Forest sector mitigation options are summarized below.

11.2.1 Maintaining Existing Stocks

- **Forest Protection and Conservation**. This protects the carbon and other GHGs in both the vegetation and soil. Such measures are often included in projects which are put in place for non-carbon resource management purposes, such as wildlife protection, soil conservation, water catchment, and recreational reserves. Measures to improve wildfire protection and reduce forest losses from insects and diseases should also be considered under this category.

- **Increased Efficiency in Forest Management, Harvesting, and Product Utilization**. Measures to increase efficiency include natural forest management with selective harvesting; harvesting for multiple end-uses; residue utilization for fuel and tertiary products; increased conversion efficiency, possibly involving technological intervention; and salvage operations during conversion of forests to other land uses like hydropower development, etc.

- **Bio-energy Initiatives**. Mitigation options related to bio-energy will mainly reduce the use of biomass and thus maintain stocks of carbon while restraining emissions of trace GHGs. According to the *IPCC 1994 Draft Guidelines for National Greenhouse Gas Inventories*, all net emissions from biomass burning should be considered loss of forest stocks. Bio-energy options include: more efficient kilns for charcoal production, charcoal packaging (e.g., briquettes), improved woodfuel stoves; improved use of charcoal for industry (e.g., steel); and more efficient use of wood in agriculture (e.g., tobacco and tea curing).

  Use of sustainably-grown biomass for fossil fuel substitution is another option. Such biomass may also be used as a substitute for fuelwood from natural forests that are being depleted.

  Urban tree planting to reduce energy use for building cooling and heating should be handled in the energy sector analysis to the extent that one can compute the energy savings.

11.2.2 Expanding Carbon Sinks

Each of the options in this category must be separately identified and described depending on the intended use of the new biomass or the fate of the new land use. Uses include provisions of forest products such as woodfuel, timber, pulp and papers, and forest services like recreation, soil protection, and emission reduction through fossil fuel substitution. The fate of the biomass is critical in determining the carbon flows, cost and benefit streams, and the implementation possibilities of the specific mitigation options listed below.

- **Afforestation**: Planting forests on bare land, with biomass density commensurate to the objective of the project

- **Reforestation**: Replanting and/or natural regeneration of deforested areas

- **Enhanced Regeneration**: Increasing the biomass density of existing degraded and under-stocked forests

- **Agroforestry:** Some or all of the agroforestry forms listed below may be applicable to different sites in a country. The most commonly practiced forms are:

  - inter-cropping for the purpose of producing both agricultural and forest products

  - boundary and contour planting for wind and soil protection, as well as for providing agricultural and wood products.

- **Urban and Community Forestry:** This includes the additional biomass in non-contiguous tree cover which has not been described elsewhere. This may include residential shade trees, and road-side and demarcation trees in the rural areas. Expanded urban forestry, which sequesters carbon and may also reduce emissions through cooling and heating of urban residential and commercial buildings, should also be considered.

11.3 COMPREHENSIVE MITIGATION ANALYSIS PROCESS (COMAP): OVERVIEW

COMAP is intended to guide an analyst in undertaking a comprehensive assessment of the role of the forest sector in a country's climate change mitigation effort. This approach includes the following specific steps (See Figure 11-1):

- Screening to identify mitigation options significant to the country
- Assessing of the current and future land area available for restraining emissions and/or carbon sequestration given the demand for land by all sectors
- Identifying the mitigation options which could be implemented on the various available lands
- Estimating of the emission reduction and/or carbon sequestration per unit area for each mitigation option
- Estimating of the total and unit costs and benefits for each option
- Developing of future GHG net emissions and cost scenarios
- Evaluating the cost-effectiveness of mitigation options
- Exploring the policies, institutional arrangements, and incentives necessary for the implementation of mitigation options.

The first step involves categorizing and screening the mitigation options that are suitable for implementation. This is followed by a determination of the forest and agricultural land area that might be available to meet future demand for both domestic consumption and export. Demand for wood products includes demand for fuelwood, industrial wood products, and construction timber. The land which is left over after satisfying the future demands for products and the demand for conversion to other land uses can be considered to be available for carbon sequestration and/or other environmental purposes. In many countries there may not be enough land available. In this case, some of the wood demand may have to be met through wood imports or by using substitutes for forest products. Alternative combinations of land use and wood product demand patterns will lead to different scenarios of the future. The most-likely-trends scenario is chosen as the base line scenario against which the mitigation scenarios are compared.

The mitigation options are then matched with the types of future wood products that will be demanded and with the type of land that will be available. Matching options and demand requires iterating between satisfying the demand for wood products and land availability considerations. Based on this information, the potential for emission reduction and/or carbon sequestration, as well as the flow of costs and benefits per hectare of each mitigation option, are determined. The carbon and cost-benefit information is used to establish the cost-effectiveness of each option, which yields its ranking among other options. In addition, this information, in combination with land-use scenarios, is used to estimate the average and total cost of each and all mitigation options. Finally, the barriers, and the policies and incentives needed for the implementation of each scenario, are explored.

Such a comprehensive approach should result in a mix of mitigation options which use the fewest resources as well as provide the most benefits while mitigating climate change. This will allow for a cost-effective implementation of a subset of the options depending on the available resources at any point in time. This approach also reduces the possibility of double-counting of GHG flows, costs, and benefits.

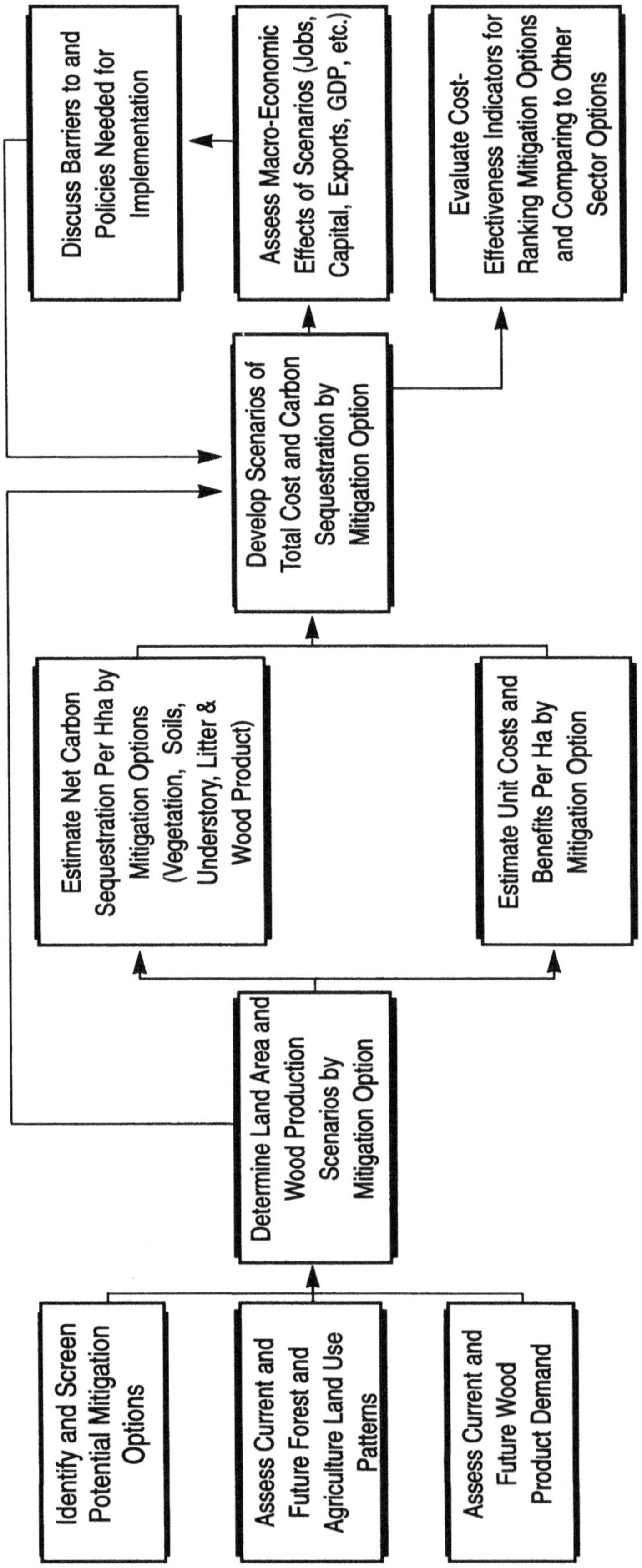

Figure 11–1: Comprehensive Mitigation Analysis Process (COMAP) for Forestry

In Appendix 11-1, two examples are presented to illustrate the comprehensive assessment approach. The first example, forest protection, is drawn from the category of mitigation options intended to maintain existing stocks of biomass and carbon. The second example, reforestation, aims to expand carbon stock. In both examples, there is also a possibility of avoiding emissions of trace GHGs.

11.4 SCREENING OF MITIGATION OPTIONS

There is a need to screen out non-promising options prior to undertaking a comprehensive evaluation. While the general criteria for screening are similar to those described in Chapter 2, there are some criteria which are specific to the forestry sector. These criteria may include: conformity with existing forest management plans, equity and co-benefits issues, feasibility and/or ease of implementation, ecological soundness of the option, etc. The following are two examples of screening criteria:

- **Biophysical considerations**. Some options may be screened out due to biological or physiographic reasons. These may include site characteristics, e.g., climate, soil, drainage, and altitude. For example, large increases in productivity in a dry area through short rotation forestry in an area without possibility of irrigation can be screened out at this stage.

- **Political considerations**. Those options which are expected to significantly infringe on the sovereignty of a country or might tend to cause political instability should be screened out. For example, a measure which requires physical removal of large numbers of forest dwellers for re-settlement may be politically infeasible and socially unwise for the country in question.

11.5. LAND AVAILABILITY AND PRODUCT DEMAND/SUPPLY

As discussed in Chapter 10, both baseline and mitigation scenarios depend on the demands exerted on the forest resource for both wood products and other land uses. However, in many countries the dynamics of the economy shape the scarcity of land. Whether available lands are ever used for biomass growth depends on economic, political, demographic, social, cultural, and other factors.

Estimation of available land area can be done as described in Chapter 10. The easiest method is to use the existing plans on land-use management, such as government plans to increase the current forest cover to a given proportion of the land area in the future. A more involved method requires matching each mitigation option with the land available for its implementation, adjusting for any possible overlap where more than one option takes place on the same piece of land. This must take into account the minimum land requirements by the other sectors over time, especially the agricultural sector which has traditionally been given priority over forestry in land-use allocation.

11.5.1 Land-use Scenarios

An important element of the approach is developing scenarios for demand on land use, wood and other relevant products. These scenarios depict the amount of products that would be demanded and the land that would be required to support such a demand, given the capacities of the various land categories. The amount of carbon that can be potentially stored, and the net cost of doing so, varies with the type of options that will be included in the scenarios. The development of scenarios, discussed below, would form the basis for applying the method described in Chapter 10.

11.5.1.1 Baseline or likely trends scenario

This type of scenario is based on the current trends of land use and consumption of forest products in the country. It involves describing existing land-use distribution among and within sectors, the rate at which land is being converted from one use to another, and identifying the factors which drive the land-use distribution and conversion process. Factors such as population and income growth rates are used to extrapolate the demand for land and forest products to the future under given assumptions on the behavior of the factors.

11.5.1.2 Mitigation scenarios

- **Technical Potential Scenario**

  This scenario helps to estimate the amount of carbon that might be stored if the technically available land area were to be fully utilized for carbon sequestration. This scenario ignores many factors—economic, institutional, cultural, legal, etc.—that may limit the usability of available land for the sole purpose of storing carbon. Thus, the scenario represents an upper limit to the amount of carbon that might be stored through forestry options in a country.

- **Programmatic Scenario**

  A programmatic scenario is one which is based on specific existing programs. Examples of such programs include the America the Beautiful reforestation program (Andrasko *et al.*, 1991), the goal declared by the Noordwijk Convention to increase net world forests by 12 million hectares a year by the beginning of the next century, and various national plans/programs and bilateral initiatives, like the Tropical Forestry Action Plan (TFAP), which are in place in many tropical countries. If one does not use the comprehensive approach, a programmatic scenario is the most appropriate type of mitigation scenario to use.

  One disadvantage of a programmatic scenario is that it may yield wood far in excess of its domestic and/or export demand. This could lower the price of wood and reduce an option's net monetary benefit, depending on the price elasticity of demand of the relevant products. Programmatic scenarios are particularly likely to create an inequitable distribution of benefits (Adams *et al.*, 1993) since they are usually driven by a single major purpose, e.g., to store carbon, to conserve fauna, to rehabilitate degraded lands, etc. However, the excess wood derived from such scenarios might be absorbed by an expanded program of using biomass to replace fossil fuels.

- **End-use Scenario**

  This type of scenario would be driven by the need for wood products and various land uses in a country. The end-use approach has been used extensively to understand the magnitude of future demand for energy (Sathaye *et al.*, 1989). However, with the possible exception of some applications in developed countries (Adams and Haynes, 1980), this approach has not been applied to the forest sector with the same analytical rigor and specificity. The Food and Agricultural Organization (FAO) has done numerous studies on timber trends and outlook both at the global and regional level, but very few such studies for individual developing countries have been done (FAO 1963, 1967a, 1967b, 1986.).

End-use scenarios have the advantage that they take into consideration an end-user's needs for forest products and land. In tropical countries, where wood is scarce and forests are used as sources of many non-timber products, planting trees for carbon storage alone may not be sustainable. The trees will most likely be cut down and used for various products. Thus, only trees that provide multiple and adequate benefits, including carbon storage, to a diverse set of beneficiaries are likely to be managed sustainably. In order to satisfy the assumption that tree stock should be maintained in perpetuity, it is important that all participants be adequately compensated. An end-use-based approach that explicitly recognizes the needs of the participants will yield more plausible and sustainable future scenarios than a programmatic approach.

Construction of end-use scenarios is done with varying degrees of complexity. The simpler forms are used to project current consumption per capita into the future, adjusting for driving factors such as population growth and income. To improve on this, one can make further adjustments using known or estimated income elasticities of demand for the product in question. Many of the wood products outlook studies mentioned above are based on this approach. For example, the approach was recently applied to evaluate forest sector mitigation options for India (Kadekodi and Ravindranath, 1994).

The second way to construct end-use scenarios involves statistical estimation of a product consumption function with a few explanatory variables used to generate the coefficients needed for projection. Time series or cross-sectional information is used depending on the product in question and availability of data (FAO, 1991). A more rigorous variation of the statistical approach involves an econometric analysis of the product market (both demand and supply), including the use of some form of a land-use allocation model for tracking the required forest areas needed to meet such demands (Adams and Haynes, 1980). This approach usually includes a policy simulation phase which is used to forecast the impact of various policies, including those which may constitute mitigation options. This method will most likely lead to more precise projections of future demand and supply of forest products and forest land. However, its application requires a sound knowledge of econometric techniques and a good amount of data on production, consumption and price structure of forest products, and applicable factors (of production) and technology. Of the three methods, each country should use the one commensurate to its capability and availability of data and be aware of the different levels of precision associated with the resulting projections.

- **Achievable Scenario**

The end-use scenario described above shows the projected demand for both forest products and land use under the various mitigation options. This scenario may not be achievable depending on the likelihood of implementing the underlying mitigation options, which depends on factors such as land tenure and law, and available technical and human resources. In most cases, the achievable scenario is less than the economically defined scenario. Past experience with implementing similar projects can serve as a guide for estimating the magnitude of GHG mitigation that may be achievable.

In summary, one should construct a baseline scenario for the purpose of computing the change caused by mitigation options. Of the four mitigation scenarios, the technical potential is of some interest in that it shows the maximum physical potential. The programmatic scenario will be useful for those countries which do not apply the comprehensive approach. Wherever possible, the analyst should construct end-use scenarios and also estimate the achievable potential.

11.6 ANALYSIS OF MITIGATION OPTIONS

The aforementioned mitigation options either maintain or expand the stock of carbon in biomass, soil, and/or wood products. Two approaches have been used in the past to evaluate stored carbon. One approach assumes that trees will be planted for the purpose of storing carbon and will not be harvested after they grow to maturity. It suggests that carbon stock be estimated according to the amount accumulated in tree biomass, soil, litter, and understory over a period of time (Moulton and Richards, 1990). The time period may be that of a single rotation or multiple finite rotations. The second approach assumes that carbon will need to be stored in perpetuity and estimates the amount of stored carbon based on an average amount of on-site carbon over an infinite number of rotations (Dixon *et al.*, 1991).

A modified version of the second approach has been used by Swisher (1991). The method adjusts average stock for the biomass remaining at maturity. Swisher also includes the carbon in soil, litter, understory, and wood products in estimating the total carbon storage of a given site.

Of the two methods mentioned above for evaluating stored carbon, the second approach is preferable, since (1) it is consistent with the fact that the economic value of stored carbon is unknown and thus the safest course of action would be to store carbon in perpetuity, and (2) intermittent harvesting of forest products provides a periodic income stream that strongly influences the financial evaluation of a mitigation option (Winjum and Lewis, 1993). See Appendix 11-2.

11.6.1 Greenhouse Gas Flows

Numerous mitigation options can be identified for implementation in different parts of the country. Their implementation will take place over a long period of time. The estimation of GHG flows should cover the following items:

1. **Greenhouse Gas Emissions**

   Estimate carbon flows associated with each option at all stages of the project, including emissions from salvage, silvicultural operations, harvesting, and short-term products like pulp and paper and woodfuel. Also to be included are the emissions from soil disturbance and from future decomposition from medium- and long-term forest products in use. The other trace GHGs, such as methane, N_2O, NO_x, and NMHC, should be estimated if the option in question will eliminate the emission of significant quantities of them into the atmosphere. Despite the existence of a substantial level of uncertainty associated with estimating emissions of these trace gases, one should, at the very least, use simple emission ratios of these gases per unit of emitted CO_2-C. Appendix 11-3 provides applicable ratios for trace gas emissions from biomass burning and flooding.

2. **Carbon Sequestration**

   This includes estimates of uptake by vegetation and soils within the area. This should be based on Net Primary Productivity of the woody biomass, including the net storage in soil and detrital material. Alternatively, carbon sequestration can be estimated for the various mitigation options based on the mean annual increment of the growing vegetation and the refurbishment of the soil carbon from organic matter.

11.6.1.1 Incremental carbon storage

In order to evaluate the incremental carbon benefit of a mitigation option, it is necessary to estimate the carbon that might have been stored without the project. For forest protection, the baseline amount of carbon stored may be estimated as that which would have been released in the absence of a protection measure such as a physical barrier, or monitoring and policing, relocation of forest dwellers, provision of alternative means of earning a living to the current users of the forest, or a subset of these measures (Swisher, 1991).

In the case of plantations or management of forests under rotation, the situation is more complicated. One needs to compare the incremental carbon which would be sequestered on land and in products indefinitely. On land, carbon will be stored in vegetation, soil, and the decomposing biomass. The carbon stored per hectare in a plantation or forest managed sustainably in rotations can be estimated as described in Appendix 1, Example 2.

It is important to include carbon stored in wood products when estimating the total carbon storage since wood-product carbon can amount to 30-40% of the carbon stored on land (Dewar and Cannell, 1992). By not including product carbon, studies such as Dixon *et al.* (1991) may overestimate the unit costs by a corresponding proportion.

A more comprehensive framework for tracking the carbon flows from all forestry mitigation options in a country is a spreadsheet model named COPATH (Makundi *et al.*, 1991). This model provides a coherent framework for tracking all the carbon flows from each scenario over as long a period as desired while providing the analyst the capability to assess the carbon balance in the sector at any point in time under each scenario. In brief, the model allows estimation of the existing carbon stock in the forest areas, which may be affected by the mitigation options, as well as estimation of the carbon release from decomposition and oxidation of the affected biomass. It also helps estimate the carbon sequestration in vegetation, soil, and wood products.

11.6.2 Estimating Costs and Benefits of Mitigation Options

To evaluate the mitigation options, a set of criteria should be assembled for each option. The evaluation should cover a variety of criteria, including physical, socio-economic, and other environmental factors so as to make comparison of options exhaustive. The physical criteria such as land availability, biomass productivity, and net GHG flows for each option have been described above. Below is a discussion on evaluating the suggested economic criteria.

- **Physical Inputs and Outputs**

  Although the physical criteria determine the capacity of the various options to mitigate climate change, the economic criteria are essential decision variables in choosing mitigation options. In this section, we discuss the approach and issues involved in compiling the economic criteria.

  The first step is to identify and quantify all necessary physical inputs required for implementing each option covering the initial operations, management, harvesting (if applicable), etc. These should include estimates of land, labor, equipment, and material needed to support the project or option throughout its lifetime. For the land-intensive options, different categories of land must be identified and their suitability for various options assessed. For all options, the analyst must also identify constraining factors such as expertise, technology, and capital investment because these may affect the cost as well as the likelihood of implementing the option.

Together with the physical inputs, one must estimate the physical output in terms of desirable products like timber, woodfuel, and agricultural produce (for agroforestry options) that are expected from each mitigation option.

- **Unit Costs and Benefits**

In order to create policies and measures to stabilize future GHG emissions, national policy-makers need information on the costs and benefits of options in addition to their carbon implications. Policy-makers must weigh the costs, benefits, and impacts of climate change mitigation and adaptation options in the face of competition for limited funding from the government and other sources. The policy goal for climate change mitigation options is to identify which mix of options will best achieve the desired forest resource utilization goals at the least cost. In other words, the policy should attempt to maximize economic and social benefits from forestry while minimizing local and global environmental and social impacts.

It is important to draw a system boundary within which the costs and benefits of a project will be evaluated. Costs and benefits should be evaluated up to the roadside and not to the millsite or market place. Roadside costs would include the cost of harvesting wood, which in turn includes the required forest-road construction costs. By choosing to report costs only up to the roadside, we exclude the costs and carbon emissions associated with transporting the produce to the market. This also eliminates the need to collect data and make projections on the location of mills which will likely change if a large magnitude of projects have to be implemented in order to significantly reduce nationwide emissions. The post-roadside costs should be handled in the respective end-use sectors such as industry (sawmilling) and residential (biomass fuels).

For each of the physical inputs (e.g., labor), one has to obtain the cost per unit at the time of use. For each desirable product (e.g., timber or woodfuel), an estimate of product price will be necessary. These will be used to calculate the cost and benefits of the monetary elements of each option.

Costs. The costs of carbon storage of a mitigation option include the (1) present value of the stream of expenses sufficient to cover the project's planning, development, and occasional and recurrent expenses, and (2) present value of the project's opportunity cost. Swisher and Masters (1992) refer to the present value of future project costs as an endowment.

The endowment includes the initial cost of establishing the project, cost of silvicultural operations, management, extension services, protection, and cost of monitoring the project's performance. For perpetual management of a given forest project, the benefits derived during the first rotation may be sufficient to cover the operation and management of future rotations.

The 1990 IPCC report on Response Strategies to Climate Change reviewed the then existing literature on costs and benefits and noted that halting deforestation was a low-cost option for reducing a unit of atmospheric carbon (IPCC, 1991). The report quoted regional average annual costs of about $8/tC for tropical forestation and reduction of deforestation, and about $28/tC for forestation in non-US OECD countries. The cost of establishing a forest plantation, excluding the opportunity cost of land, was estimated to range from $230 to $1000 per hectare with an average cost of $400 per ha (Sedjo and Solomon, 1988).

The unit cost estimates for carbon sequestration have been improved in several ways since the IPCC report. First, unit costs have been estimated for individual countries by different types of mitigation options rather than by regions or for the globe as a whole. For example, Dixon *et al.* (1991) estimated establishment costs for 94 countries by combining survey data with information gathered from the literature.

Secondly, other cost components such as land rental (opportunity costs), maintenance, and monitoring and evaluation, which were not included in the earlier IPCC report, are now being addressed (Swisher, 1991; Moulton and Richards, 1990). The opportunity cost evaluation is important since it captures the benefits derived from land use in the absence of a mitigation option, given the current broad land-use patterns. Opportunity cost may be evaluated using various methods depending on the land in question and the likelihood of producing various goods and/or services if it is not used for the given option. These approaches include land rent, land market price, and net benefits obtainable from an alternative land use. In all these cases, land values and benefits from alternative use should be adjusted to account for existing significant price distortions due to subsidies, zoning regulations, etc. For example, land rental costs for the US were estimated at $142 per ha by Moulton and Richards (1990), and the land purchase price was estimated between $400 and $1000 per ha by Sedjo and Solomon (1988).

Land prices are likely to be lower in developing countries. For Thailand, Wangwacharakul and Bowonwiwat (1994) reported an estimate of $44-89 per ha for present value of the opportunity cost of land. For degraded lands suitable for reforestation in India, the land price is very low ($20/ha) (Ravindranath and Somashekhar, 1994). For China, the forest lands are already allocated for forest development, while the dry croplands are reserved for agroforestry development. Thus, the opportunity cost of non-forestry land use or land classified for forestry may be close to zero (Xu, 1994).

Benefits. In addition to carbon storage, implementing a mitigation option will result in other monetary and non-monetary benefits. These benefits may be classified as direct or indirect benefits depending on their role in, and level of, economic activity and non-monetary forest values. Direct benefits may include goods such as fuelwood and timber and services such as recreation. Indirect benefits may include such items as employment for local inhabitants, air pollution and micro-climate control, watershed protection, and development of social infrastructure such as schools, roads, and hospitals. In addition to these benefits, the forest has a value derived from the stock as a resource. This value may be influenced by concern for future generations and social status.

There is no consensus at present on the monetary value of reducing a unit of atmospheric carbon. Preliminary estimates of the marginal cost (including taxes) of stabilizing emissions from fossil-fuel burning in the US range between $100 to $200 per tC (Cline, 1992; Nordhaus, 1993), based on top-down models which do not include significant improvements in low-cost energy intensity in the economy. The unit cost estimates for mitigation options in most of the F7 countries[1] fall well below this range, and in the case of India they are also below the unit costs of energy-efficiency options.

---

[1] The F7 network is a group of researchers from nine countries who have been doing research on Tropical Forestry and Global Climate Change, coordinated by Lawrence Berkeley Laboratory, for the past four years. The countries are Brazil, China, India, Indonesia, Malaysia, Mexico, Nigeria, Tanzania, and Thailand.

A unique feature of the methodology presented in this section is the explicit evaluation of the direct benefits which may be derived from the sale of timber and other wood and non-wood products. As has been demonstrated in the F7 studies (Sathaye *et al.*, 1993), the benefits are sufficiently large to offset the life-cycle cost of many sink expansion options. In effect, carbon may be sequestered at a net benefit to society.

11.7 COST-EFFECTIVENESS EVALUATION

11.7.1 Initial Cost per ha and per tC

Initial costs do not include future discounted investments needed during the rotation period. This indicator provides useful information on the amount of resources required to establish the project.

Most cost studies estimate this indicator (Dixon *et al.* 1991; Andrasko *et al.* 1991; Volz *et al.* 1991). The other cost components and the option's benefits are often ignored. The studies take into consideration the carbon stored in live biomass and most account for soil carbon. The Dixon (1991) study uses the mean stock of carbon as a measure of the amount of carbon that would be stored by a mitigation option. The other studies report several estimates of the cost per tC, but their method of carbon estimation is unclear.

11.7.2 Endowment Requirements per ha and per tC

This is the sum of establishment cost and the discounted value of all future investment and recurring costs during the lifetime of the project. For rotation projects, the costs of second and subsequent rotations would be paid for by the revenues derived from the first rotation and thus would not be included in estimating the endowment. For projects which do not have substantial monetary benefits, this indicator is quite useful because it provides the endowment necessary to maintain the project in perpetuity. Swisher (1991) uses this indicator to evaluate project cost-effectiveness.

11.7.3 Net Present Value (NPV) per ha and per tC

This indicator would provide the net direct benefit to be obtained from the project. For most plantations and managed forests, it should be positive at a reasonable discount rate. For options such as forest protection and reforestation (Appendix 11-1), the NPV indicator is also positive if indirect benefits and forest value are included, both of which are subject to controversial evaluation. Appendix 11-2 provides the mathematical formulation for deriving this indicator for plantations and managed forests.

11.7.4 Benefit of Reducing Atmospheric Carbon (BRAC)

This indicator expresses the NPV of a project per unit of atmospheric carbon reduced as opposed to the reduction of net emissions. In so doing, it captures the atmospheric residence time of carbon. The formulation of the indicator varies with the rate at which economic damage might increase, and it allows time-dependent evaluation of atmospheric carbon as may be deemed necessary. The expression for deriving BRAC when the economic damage caused by atmospheric carbon increases at the real societal rate of discount is given below. For complete coverage of the BRAC indicator, see Sathaye *et al.* (1993).

$$BRAC = NPV \,/\, \left(\frac{1}{a} \sum_{0}^{Te} C_t\right)$$

where NPV = Net present value of benefits
a = Decay rate of carbon
T_e = Time duration of carbon flows
C_t = Net carbon flow in time t

11.7.5 Imputed and Non-monetary Costs and Benefits

After compiling the criteria given above, all the identifiable costs and benefits that one is currently unable to evaluate should be listed for each mitigation option. Imputed values should be listed separately from the direct costs and benefits. The intangible benefits and costs should also be listed in a separate column for each mitigation option. To the extent it is possible, identify the likely bearers of costs and benefits, including the non-monetary items. All these criteria may play an essential role in the choice and subsequent implementation of mitigation options.

11.8 DEFINING BASELINE AND MITIGATION SCENARIOS

Having compiled the physical, economic, and the intangible criteria for each mitigation option, one uses them to define mitigation scenarios containing a set of options. One useful method is to summarize the results as a supply curve for emission reduction or carbon sequestration. For example, the establishment cost per tonne of carbon or per hectare can be used to plot a cost of conserved carbon (CCC) curve for all mitigation options. The curve shows the amount of carbon that could be stored at increasingly higher establishment cost. The other indicators (endowment requirement, NPV, or BRAC) could also be used to plot similar curves.

The unit values are combined with the area availability for each option to obtain a step function of all options in a scenario. From these curves one can calculate the total area and financial resources required to achieve a given scenario, and the total amount of carbon and other GHGs saved. Although there may not be a coherent method for comparing the non-monetary and intangible costs and benefits associated with each option, their enumeration helps the policy-makers in the choice and implementation of the various mitigation options.

11.9 MITIGATION POLICIES

11.9.1 Identifying Implementation Policies

Having constructed the baseline and mitigation scenarios, one has to identify and describe the policies which may be necessary to implement the mitigation options. These policies can be divided into two groups: (1) forestry policies which govern the use of forest resources, and (2) non-forestry policies which happen to influence levels of activities in the forestry sector.

(1) Forestry Policies

Policies that can be used to maintain carbon stocks and/or expand carbon sinks include the following:

- Forest protection and conservation policies. One has to consider both national, regional, and local measures to preserve existing forests and vegetation cover. Examples include local laws prohibiting conversion of steep slopes to agricultural lands or national laws gazetting vulnerable ecosystems into nature reserves.

- Policies on shared responsibility for managing existing protected areas between local communities and the central agencies, which also include the sharing of benefits from the protected area, tend to reduce "encroachment" by the surrounding population. Such policies have been applied effectively in many developing countries. A recent example is the shared wildlife management in Zimbabwe.

- Policies governing terms of timber harvest concessions (allowable cut, concession duration, levels and structures of fees and royalties) will influence the implementation and effectiveness of mitigation options that improve the efficiency of forest and product utilization. These policies may even include logging bans in specified ecosystems. Policies which emphasize export of higher-value timber products or ban log exports may reduce the rate of forest degradation while maintaining the forest sector's contribution to the country's foreign exchange earnings.

- Tax rebates and dissemination policies governing the adoption of efficient charcoal kilns and wood stoves have been shown to substantially affect the success of such programs in the bioenergy field.

- Aggressive afforestation and reforestation policies both by villagers and forest departments will help expand the carbon sinks in a country. These policies may include village afforestation schemes and incentives for private ownership of degraded lands for reforestation, emphasizing expanding plantation forestry for industrial wood instead of relying on natural woodlands.

(2) Non-forestry Policies

These policies are intended to manage the other sectors of the economy, but have large influences on the depletion of the carbon stock, and at times may provide a disincentive to increasing forest cover. The mitigation policies which lie in this area are:

- Land tenure policies that encourage private ownership of some lands with an express mandate to sustainably develop them. Policies to the contrary have been shown to encourage wasteful conversion of forests to other land uses so as to meet the criteria for property rights assignment.

- Land tenure policies that increase the certainty of tenure tend to encourage the owners of the land to plant and retain trees on their land. Such policies will be necessary for mitigation options involving agroforestry or woodfuel plantations.

- Agricultural policies that do not encourage extensive and wasteful conversion of natural forests to agricultural lands. Policies which emphasize more intensive farming and conversion of fewer marginal woodlands tend to lead to production of the same agricultural output from less area using the same amount of resources.

- Infrastructural policies governing mining, dam construction, and road construction can reduce unnecessary emissions.

- Taxes, credits, and pricing policies also play an important role.

11.9.2 Barriers and Incentives for Implementation

The policies described in the last section may not easily be translated to mitigation programs due to the existence of barriers and lack of incentives to implement them. A diverse array of criteria will have to be satisfied before a project can be implemented. The analyst should identify and describe potential solutions to these barriers. The most common barriers to the implementation of forestry mitigation options can be divided into three categories:

(1) Technical and Personnel Barriers

- Availability of scientific data on silvicultural practices and soil conservation may be a limiting factor in evaluation of various options. Availability of seed material, research on species provenance, multi-cultural management, including harvesting techniques and silvi-pastoral systems, may be lacking for individual sites.

- In the short to medium term, there may be a lack of qualified local personnel to carry out the projects and provide extension services necessary for the successful involvement of local populations.

(2) Financial and Resource Barriers

- Funding for forestry projects has been very low in most cases. Participation of the commercial sector may depend on incentives for long-term investment in forestry. The borrowing rates from banks may be too high for private investors and/or local communities to get credit for forestry projects. Bilateral and foreign-source funds are restricted to those forestry sections that are more profitable, and as such there may not be enough funds for broad investment in the identified response options.

- Agricultural activity may compete for labor with the forestry sector, depending on the types of crops and the seasonal demands on labor.

- Procedures and mechanisms for identifying beneficiaries, cost-bearers, and ways to apportion credit from the options may be a barrier to implementation.

(3) Institutional and Policy Barriers

- Land tenure and land law may prove be the strongest hindrance in implementing the mitigation options, especially in developing countries.

- Institutions necessary to allow various parties to participate in the options may not exist in the country. For example, there may not be a mechanism for sharing benefits between the central authorities and the local participants in community-based mitigation options.

- Policy barriers to harvesting, marketing of forest products, pricing, tariffs, and quotas for exports and imports may also hinder implementation of some of the mitigation options.

REFERENCES

Adams, D. M. and R. W. Haynes. 1980. The 1980 Softwood Timber Assessment Market Model: Structure, Projections, and Policy Simulations. *Forest Science Monograph No. 22.*

Adams, R., Adams, D., Callaway, J., Chang, C., and B. McCarl. 1993. Sequestering Carbon on Agricultural Land: Social Cost and Impacts on Timber Markets. *Contemporary Policy Issues*, Vol. XI, Jan., pp. 76-87.

Andrasko, K., Heaton, K. and S. Winnett. 1991. "Estimating the costs of forest sector management options: overview of site, national and global analyses." In *Proceedings of a technical workshop to explore options for global forest management,* Bangkok. 22 pp.

Aselmann, I. and P.J. Crutzen. (1990). A global inventory of wetland distribution and seasonality, net productivity, and estimated methane emissions. In *Soils and the Greenhouse Effect*, A.F. Bouman (ed.), Wiley, New York, New York., pp. 441-449, 575 pp.

Cline W. (1992). "The Economics of Global Warming." Institute for International Economics, Washington, DC.

Crutzen, P.J. and Andreae, M.O. (1990). "Biomass burning in the tropics: Impact on atmospheric chemistry and biogeochemical cycles." *Science*, Vol. 250, pp. 1669-1678.

Dewar, R. and M. Cannell. (1992). "Carbon sequestration in the trees, products and soils of forest plantations: an analysis using UK examples." *Tree Physiology*, Vol.11, pp. 49-71.

Dixon, R., Schroeder, P. and J. Winjum (1991), *Assessment of Promising Forest Management Practices and Technologies for Enhancing the Conservation and Sequestration of Atmospheric Carbon and Their Costs at the Site Level*, UE EPA/600/3-91/067, p. 24.

Food and Agricultural Organization (FAO) (1963). Latin American Timber Trends and Prospects, UN/FAO, Rome.

FAO (1967a). Timber Trends and Prospects in Africa, UN/FAO, Rome.

FAO (1967b). World Wood Trends and Prospects. UN/FAO, Rome.

FAO (1986). Forest Products World Outlook Projections 1985-2000, UN/FAO, Rome.

FAO (1991). Forest Products World Outlook Projections: Projection of Consumption and Production of Wood-based Products to 2010. No 84. UN/FAO, Rome.

Hall, D., Mynick, H. and R. Williams. (1991). Cooling the Greenhouse with Bioenergy. *Nature*, Vol.353, Sept. 5, pp. 11-12.

Intergovernmental Panel on Climate Change (IPCC) (1991). "Climate Change: The IPCC Response Strategies," Island Press, Washington DC.

IPPC (1994). "Greenhouse Gas Inventory Reporting Instructions." IPCC Draft Guidelines for National Greenhouse Gas Inventories. Vols. I and II. IPCC/OECD.

Kadekodi, G.K. and N.H. Ravindranath (1994) "Economic and policy analysis of role of forestry as a climate mitigation option for India." Lawrence Berkeley Laboratory Draft Report, Berkeley.

Makundi, W. R., J. A. Sathaye and A. Ketoff. 1991. COPATH: Description of a Spreadsheet Model for Estimation of Carbon Flows Associated with Forest Use. Lawrence Berkeley Laboratory Report, LBL-30525. Berkeley.

Moulton, R.J. and K.R. Richards (1990) *Costs of sequestering carbon through tree planting and forest management in the United States.* USDA, Forest Service, General Technical Report WO-58. 48 pp., December.

Nordhaus W. (1993). Climate and Economic Development: Climates Past and Climate Change Future. Annual Bank Conference on Development Economics, The World Bank, Washington DC.

Ravindranath, N.H. and B.S. Somashekhar (1994) "Potential and economics of forestry options for carbon sequestration in India." *Biomass & Bioenergy*, Special Issue (in press).

Sathaye, J., Ketoff A., Schipper, L. and S. Lele (1989). "An End-Use Approach to Development of Long-term Energy Demand Scenarios for Developing Countries." Lawrence Berkeley Laboratory Report, LBL-25611, Berkeley.

Sathaye, J., Norgaard, R. and W. Makundi. (1993). "A Conceptual Framework for the Evaluation of Cost Effectiveness of Projects to Reduce GHG Emissions and Sequester Carbon," Lawrence Berkeley Laboratory Report, LBL-33859, Berkeley.

Sedjo, R. and A. Solomon. (1988). Climate and Forests. In N. Rosenberg, *et al.*, Eds. "Greenhouse Warming: Abatement and Adaptation," 105-19, Washington: Resource for the Future.

Swisher, J.N. (1991) "Cost and performance of CO_2 storage in forestry projects." *Biomass and Bioenergy* 1(6):317-328.

Swisher, J. and G. Masters (1992) "A mechanism to reconcile equity and efficiency in global climate protection: international carbon emission offsets." *Ambio* 21(2):154-159.

Volz, H., Krriebitzsch, W. and T. Schnieder. (1991). "Assessment of Potential, Feasibility and Costs of Forestry Options in the Temperate and Boreal Zones," in *Proceedings of Technical Workshop to Explore Options for Global Forestry Management,* Bangkok.

Wangwacharakul, V. and R. Bowonwiwat (1994) "Economic evaluation of CO_2 response options in forestry sector: Thailand." *Biomass & Bioenergy*, Special Issue (in press).

Winjum, J.K. and D.K. Lewis. (1993). "Forest management and the economics of carbon storage: the nonfinancial component." *Climate Research*: 3:111-119.

Xu, D. (1994) "The potentiality of reducing carbon accumulation in the atmosphere by large-scale afforestation in China and related cost/benefit analysis." *Biomass & Bioenergy*, Special Issue (in press).

APPENDIX 11-1
EXAMPLES OF COMPREHENSIVE ASSESSMENT APPROACHES FOR THE FORESTRY SECTOR

EXAMPLE 1: MAINTAIN CURRENT CARBON STOCKS – FOREST PROTECTION

The steps in this example are illustrated in Table 11-1 for the hypothetical Country X.

Steps 2 and 3: Determine current and future land area under two scenarios, baseline and mitigation options. In our example, the baseline scenario assumes that in 1990 there were 12,000 ha of protected forests which had declined from 15,000 ha in 1980. It is further assumed that the forest degradation and conversion of land to agriculture would continue in the future until only 1000 ha are left in the protected area by 2030. Conversion of land to annual agriculture implies that the stored carbon is reduced from a high equilibrium consistent with forest ecosystem to a much lower level consistent with annual crop production.

In the mitigation scenario, adequate steps are taken to ensure that the area is protected and the 12,000 ha of land remain protected until 2030. In addition, protection increases the biomass density and carbon density of the protected area.

Step 4.1: Determine the current and future biomass density under each scenario. In order to determine the carbon pool and sequestration, it is necessary to know the biomass density, the soil carbon density, and the carbon content of biomass. If this information is not available from destructive sampling data, it can be estimated using the following formula:

$$\text{Dry Biomass Density}(t/ha) = SV * AS * TA * DW * WD$$

where SV = Stemwood Volume (m^3/ha)
AS = Above-ground biomass over Stemwood volume ratio
TA = Total biomass (above plus below-ground) to Above-ground ratio
DW = Dry to Wet biomass ratio
WD = Wood Density (t/m^3)

In our baseline scenario, we assume that the biomass density continues to decline as the forest area is degraded, starting with 200 t/ha in 1980 to 160 t/ha by 1990. The density is assumed to continue declining at a rate of 1% each year to 107 t/ha by 2030.

Alternatively, the dry biomass density increases in the mitigation scenario at 1% annually to reach 238 t/ha by 2030.

Step 4.2: Determine the current and future carbon density under each scenario. The carbon ratio of biomass varies between 0.45 and 0.55 for most vegetation, with a few exceptions like rubber, which can have substantially higher carbon content. Multiplying the biomass density by the carbon ratio (C%) yields the carbon density (tC/ha) for each scenario. In our example, we assume that the carbon ratio (C%) is the same for both baseline and mitigation scenarios at 0.5. Biomass carbon declines from 80 tC/ha to 54 in the baseline scenario and increases to 119 tC/ha by 2030 in the mitigation scenario.

Step 4.3: Determine the soil carbon density for each scenario. We assume that the soil carbon density remains unchanged at 100 tC/ha in the baseline scenario but increases with the gradual rise in biomass density to 149 tC/ha by 2030 in the mitigation scenario.

Step 4.4: Determine the total carbon density under each scenario. Adding the biomass and soil carbon density yields the total carbon density for each year under each scenario. Total carbon density (tC/ha) decreases from 180 to 154 for the baseline scenario and increases to 268 in the mitigation scenario.

Step 5.1: Determine the cost of forest protection. In our baseline scenario, the area was poorly protected and the expenditure, which may be based on the annual government budget, was $2/ha/year. In the mitigation scenario, this increases to $9/ha/yr which is assumed to provide adequate protection to the area. This figure is the present value of a stream of costs from 1991 to 2030 (**Step 5.1.1**). Initial costs are $5/ha in 1991 and recurrent costs are $0.5/ha/year until 2030.

Step 5.2: Determine the benefits from land conversion. In the baseline scenario, a portion of the protected land was lost each year to encroachment from which settlers derived monetary benefits, which we assume to be $50/ha/year. No land conversion occurs under the mitigation scenario.

Step 5.3: Determine the cost or benefit of alternative means of satisfying the settlers' demand for products and services. In the mitigation scenario, the settlers would either occupy other lands, which would have carbon consequences, or their needs would have to be met through import of products. We assume, for simplicity, that their needs would be met by imports of products. The nation would bear the cost of these imports, which would normally be higher than the settlers' benefits. We assume the imports to cost 5% more than domestic products. Had the settlers' demand been met by allocation of other land, it could have resulted in a net benefit to the economy.

Step 5.4: Determine the benefits of protection. In the baseline scenario, the protected area provided certain recreation and other benefits which were valued at $2/ha/year. These increase to $15/ha/year as mitigation reduces the degradation of the vegetation in the protected area.

Step 6.1: Determine the carbon pool and annual sequestration for each scenario. Multiplying the total carbon density (tC/ha) by the land area (ha) under each scenario yields the pool (tC) of carbon for each year. Since the carbon density and the land area decline in the baseline scenario, the carbon pool declines from 2,160,000 tC in 1990 to 153,518 in 2030. In the mitigation scenario, it increases to 3,215,946 tC by 2030.

It is important to compute the annual increase in the carbon pool in the mitigation scenario and compare it with the annual change in the baseline scenario. The annual incremental carbon is 80,480 tC in 1991, which declines to 74,748 tC by 2030.

Step 6.2: Determine the total costs and benefits. Here we aggregate the costs and benefits for the baseline and mitigation scenarios. For the baseline scenario, the cost of forest protection is computed by multiplying the **Step 5.1** cost by the protected area. For example, for 1991, the cost is calculated as $2/ha/year * 11725 ha = 23,450 $/year. Similarly, the benefits of land conversion are computed by multiplying the figures for 1991 in **Step 5.2** by converted land area, and benefits from protection are calculated by multiplying the figures in **Step 5.4** by land area. Adding these three estimates yields the net benefit for the baseline scenario for each year ($13,750 for 1991).

The costs and benefits for the mitigation scenario for 1991 amount to a net benefit of $52,343. The incremental net cost for 1991 = -$52,343 + $13,750 = -$38,593 or the incremental net benefit for 1991 = $126,300.

The present value of the stream of costs and benefits from 1991 through 2030 is computed next. In this illustration, we assume a discount rate of 10%. Experts should use discount rates appropriate to their economies. The present value of incremental net cost amounts to $1,936,317.

Step 7: The cost-effectiveness of conserving carbon may now be expressed using the aforementioned indicators. The net present value of benefits is -$0.63/tC and -$161/ha and the corresponding BRAC value is -$0.047/tC. The initial cost of forest protection is $0.02/tC and $5/ha. The endowment required to protect forests until 2030 is $0.23/tC and $59/ha.

Table 11-1
Analysis of a Forest Protection Mitigation Option, Country X

FOREST PROTECTION		1980	1990	1992	2030
>>> FROM STEPS 2 AND 3: LAND AREA (ha)					
>> Baseline Scenario		15000	12000	11450	1000
> Land Converted from Forest				275	
>> Mitigation Scenario		15000	12000	12000	12000
>>>> STEP 4: ESTIMATING CARBON POOL AND SEQUESTRATION					
>>> STEP 4.1: BIOMASS DENSITY (t/ha)					
>> Baseline Scenario		200	160	157	107
>> Mitigation Scenario		200	160	163	238
>>> STEP 4.2: BIOMASS CARBON DENSITY (tC/ha)	< C(%) >				
>> Baseline Sc	0.5	100	80	78	54
>> Mitigation	0.5		80	82	119
>>> STEP 4.3: SOIL CARBON DENSITY (tC/ha)					
>> Baseline Scenario		100	100	100	100
>> Mitigation Scenario		100	100	102	149
>>> STEP 4.4: TOTAL CARBON DENSITY (tC/ha)					
>> Baseline Scenario		200	180	178	154
>> Mitigation Scenario			180	184	268
>>>> STEP 5: ESTIMATING COSTS AND BENEFITS					
>>> STEP 5.1: COST OF FOREST PROTECTION ($/ha/yr)					
>> Baseline Scenario		2	2	2	2
>> Mitigation Scenario			9	9	9
>>> STEP 5.1.1: STREAM OF COSTS AND PRESENT VALUE ($/ha.)					
>> Initial Costs					
>> Recurrent (Maintenance etc.) Costs				0.5	0.5
>> Monitoring Costs					
>> Total Costs				0.5	0.5
>> Present Value of Costs			9		
>>> STEP 5.2: BENEFIT FROM LAND CONVERSION ($/ha/yr)					
>> Baseline Scenario		50	50	50	50
>>> STEP 5.3: BENEFIT OR COST OF PROVIDING ALTERNATIVE PRODUCTS ($/yr)					
>> Mitigation Scenario				-28875	-577500
>>> STEP 5.4: BENEFIT FROM FOREST PROTECTION ($/ha/yr)					
>> Baseline Scenario		2	2	2	2
>> Mitigation Scenario			15	15	15

Table 11-1,cont'd

Analysis of a Forest Protection Mitigation Option, Country X

FOREST PROTECTION	1980	1990	1992	2030	
>>> STEP 6.1: TOTAL CARBON POOL (tC)					< Total >
>> Annual Incremental C Protected			80164	74748	3062428
>> Baseline Scenario C	3000000	2160000	2042772	153518	153518
>> Mitigation Scenario C Pool		2160000	2203416	3215946	3215946
>>> STEP 6.2: TOTAL COSTS AND BENEFITS OF CSEQ ($)					<Present Value ($) >
>> Incremental Net Cost			-10405	1060720	2129949
>> Baseline Scenario Benefit			27500	550000	1357559
> Cost			22900	2000	180395
> Benefit from Land Conversion (Opportun			27500	550000	1357559
> Benefit from Forest			22900	2000	180395
>> Mitigation Scenario Benefit			37905	-510720	-772390
> Cost			113220	113220	1107182
> Alernative Supply of Imported Products			28875	577500	1425437
> Benefit			180000	180000	1760229
>>> STEP 7: COST-EFFECTIVENESS INDICATORS					
>> Net Present Value of Benefits					
> $/tC					-0.70
> $/ha.					-177
>> Benefit of Reducing Atmospheric Carbon					
> $/tC-yr.					-0.052
>> Initial Cost of Forest Protection					
> $/tC					0.020
> $/ha.					5
>> Endowment (Net Present Value of Costs)					
> $/tC					0.25
> $/ha.					64

EXAMPLE 2: EXPAND CARBON STOCKS – REFORESTATION

The steps in this example are illustrated in Table 11-2 for the hypothetical Country X.

Steps 2 and 3: Determine current and future land area for the baseline and mitigation options scenarios. In this example, it is assumed that for the baseline scenario 40,000 ha of wasteland exist in 1990 and persist at its current level until 2030.

In the mitigation scenario, adequate steps are taken to ensure that 1000 ha are reforested each year so that by 2030 the wasteland is converted to closed forest. The reforested land will be managed in rotations consistent with the planted species and the desired forest products.

Step 4: Determine the current and future carbon pool, emissions, and sequestration for each scenario.

Step 4.1: Baseline Scenario: In order to determine the carbon pool and sequestration of wastelands, it is necessary to estimate the 1) biomass density, 2) carbon content of biomass, and 3) soil carbon density. The dry biomass density (t/ha) may be expressed as

$$\textit{Dry Biomass Density}(t/ha) = SV * WD * TA * DW * AS$$

where SV = Stemwood Volume (m^3/ha)
AS = Above-ground biomass over Stemwood volume ratio
TA = Total biomass (above plus below-ground) to Above-ground ratio
DW = Dry to Wet biomass ratio
WD = Wood density (t/m^3)

In our baseline scenario, we assume that the biomass density remains fixed until 2030 at 20 t/ha. Multiplying the biomass density by the carbon ratio (C%) yields the carbon density (tC/ha) for each scenario. We assume a carbon ratio of 45%.

The soil carbon density is assumed to be 70 tC/ha. Experts will have to obtain data for their countries in order to ascertain the density for wastelands.

Step 4.2: Mitigation Scenario: Reforestation has the potential to increase carbon density through increased carbon in vegetation, soil, decomposing matter, and wood products. The carbon density may be computed using the following procedure:

Total carbon stored = Land carbon + Product carbon
Land Carbon = (Vegetation + soil + decomposing matter) carbon

The computation of each term in the above formula for stored carbon is summarized in the equation given below. A brief description of the elements and associated assumptions is given after the equation.

$$\textit{Carbon Stored per ha} = c_v * T/2 + c_d * t/2 + c_s * T + \sum_i c_{p\,i} * n_i$$

- **Vegetation Carbon:** For the plantation response option, consider that the plantation is operated in rotation for an indefinite time period. This would ensure that at least half the carbon sequestered by an individual plot is stored away indefinitely. The formula for estimating the amount of carbon stored per ha is:

  Vegetation Carbon stored per ha = c_v*T/2
  where c_v = average annual net carbon sequestered per hectare
  T = rotation period

  In our example, we assume that the planted species has a rotation period of 10 years, a yield of 12 t/ha/year, and a carbon ratio of 0.5. Users may wish to change these values from one year to another if the species planted are different in each year or management regime is expected to vary.

- **Decomposition is equivalent to storing carbon:** The decomposing biomass on land also creates a stock of carbon. In perpetual rotation analysis, this carbon stored in the biomass may be estimated using the following formula:

  Decomposing Matter carbon stored per ha = c_d*t/2
  where c_d = average annual carbon left to decompose per hectare
  t = Decomposition period

  In our example, we assume that the decomposition period is 6 years, and the amount of decomposing carbon left behind is 6 tC/ha/year. Users should apply values applicable in their case.

- **Soil Carbon:** There is considerable uncertainty in the literature regarding the soil carbon content and the influence of factors that affect it. Hence, we should analyze economic costs and benefits with and without considering soil C. Where soil carbon data are not available, soil carbon data from other countries with similar conditions may be used. Note that the increase in soil carbon is more significant (i.e., higher percent of total carbon benefit) where the current above-ground biomass is low and vice versa. Further, we assume that the soil carbon loss and gain during harvesting and regrowth are very small compared to initial gain on degraded land.

  Soil Carbon stored per ha = c_s*T
  where c_s = Increase in soil carbon per hectare
  T = rotation period

  In our example, we assume that the soil carbon increases at 2 tC/ha/year over the rotation period of 10 years and then remains fixed in the soil in perpetuity. Users may wish to apply different values if the trend of soil replenishment is known for the area given the species.

- **Forest products:** If the forest products are renewed continually, they can store a stock of carbon over an infinite period. The amount of carbon stored in the form of products will depend on the product life. The longer the product life, the more carbon will be stored away. The amount stored over an infinite horizon, which assumes that products are replenished at the end of their lifecycle, will increase with product life according to the formula:

$$\textit{Carbon stored per ha} = \sum_i c_{p\,i} * n_i \,/2$$

where c_{pi} = amount of carbon stored per ha in product i
n_i = life of product i

We assume that the product oxidizes or decomposes at the end of its lifetime. If the character of disposal of the product is different, then applicable assumptions should be used.

In our example, we assume that the average product life is 30 years and the amount of carbon in the product is 30 tC/ha.

The total carbon stored by the mitigation option is the sum of the four components discussed above, which amounts to 583 tC/ha. The pool of carbon stored is the sum of this amount and the baseline soil carbon of 70 tC/ha, for a total pool of 653 tC/ha.

Step 4.3: We summarize the carbon density estimated in **Steps 4.1 and 4.2** for both scenarios in this step.

Step 5.1: Determine the cost of reforestation. In our baseline scenario, the cost per ha is assumed to be $5/ha. In the mitigation scenario, reforestation incurs an initial cost in 1991, 1992 and 1993. Recurrent maintenance costs are incurred which increase from $10/ha to $100/ha as the reforested area expands from 100 ha to 1000 ha. Similarly, monitoring costs increase from $5/ha to $50/ha. The stream of total costs per ha are shown in **Step 5.1.1.** The present value of these costs is $2,927/ha.

The present value of the stream of costs and benefits from 1991 through 2030 is computed using a discount rate of 10%. Users should apply a discount rate appropriate to their economy.

Step 5.2: Determine the benefits from land conversion. In the baseline scenario, the annual benefits from working the wastelands amount to $20/ha. For the mitigation scenario, the benefits are derived from timber production in the tenth and subsequent years, and from the sale of fuelwood and fruit which may be collected annually. The total benefits amount to $7.5/ha in 1991 **(see Step 5.2.1)** which then increase to reach an annual equilibrium value of $175/ha/year. The present value of these benefits amounts to $5,663/ha.

Step 6.1: Determine the carbon pool and annual sequestration for each scenario. Multiplying the total carbon density (tC/ha) by the land area (ha) under each scenario yields the pool (tC) of carbon for each year. Since the carbon density and the land area remain unchanged in the baseline scenario, the carbon pool stays at 3,160,000 tC. In the mitigation scenario, it is higher at 3,734,000 tC in 1991, which continues to increase as the fraction of land area being reforested increases.

The annual incremental carbon is 574,000 tC and the total pool is 22,960,000 tC by 2030.

Step 6.2: Determine the total costs and benefits. Here we aggregate the costs and benefits for the baseline scenario and also for the mitigation scenario. For the baseline scenario, the annual wasteland costs are $200,000, and the corresponding benefits are $800,000 for the 40,000 ha of land.

The costs and benefits for the mitigation scenario for 1991 amount to a net benefit of $3,320,830. These continue to increase as an increasing fraction of the wasteland is reforested. Net benefits reach $2,735,883 by 2030.

The difference between the baseline and mitigation scenarios' net benefits yields the total incremental benefit whose present value is $22,975,791.

Step 7: The cost-effectiveness of conserving carbon may now be expressed using the aforementioned indicators. The net present value of benefits is $1.00/tC and $574/ha, and the corresponding BRAC value is $0.075/tC. The initial cost of reforestation is $3.4/tC and $1946/ha. The endowment required to reforest and maintain the tree stands until 2030 is $1.19/tC and $684/ha.

Table 11-2
Analysis of a Reforestation Mitigation Option, Country X

REFORESTATION	1990	1992	1993	2030
>>> FROM STEPS 2 AND 3: LAND AREA (ha)				
>> Baseline Scenario				
> Wasteland	40000	40000	40000	40000
>> Mitigation Scenario				
> Wasteland	40000	38000	37000	0
> Reforested Land		1000	1000	1000
>>>> STEP 4: ESTIMATING CARBON POOL AND SEQUESTRATION				
>>> STEP 4.1: BASELINE SCENARIO -- WASTELANDS				
>> Vegetation Carbon				
> Dry Weight (t/ha)		20	20	20
> Carbon density (%)		0.45	0.45	0.45
>> Soil Carbon				
> Amount of carbon stored in soil (tC/ha		70	70	70
>>> Carbon Pool (tC/ha)		79	79	79
>>>> STEP 4.2: MITIGATION SCENARIO -- REFORESTATION				
>> 1. Vegetation Carbon		30	30	30
> Rotation Period (Years)		10	10	10
> Annual Biomass Yield (t/year/ha)		12	12	12
> Carbon density (%)		0.5	0.5	0.5
>> 2. Soil Carbon		20	20	20
> Rotation Period (Years)		10	10	10
> Amount of carbon stored in soil (tC/ha		2	2	2
>> 3. Decomposing Matter Carbon		15	15	15
> Decomposition Period (Years)		6	6	6
> Amount of decomposing carbon (tC/ha)		5	5	5
>> 4. Product Carbon		45	45	45
> Average Age (Years)		30	30	30
> Amount of carbon stored in product (tC		30	30	30
>>> Mean Carbon Storage (tC/ha)		110	110	110
>>> Carbon Pool (tC/ha)		180	180	180
>>> STEP 4.3: TOTAL CARBON DENSITY (tC/ha)				
>> Baseline Scenario				
> Wasteland		79	79	79
>> Mitigation Scenario				
> Wasteland		79	79	79
> Reforested Land		180	180	180
>>>> STEP 5: ESTIMATING COSTS AND BENEFITS				
>>> STEP 5.1: COSTS ($/ha)				
>> Baseline Scenario (Wasteland	5	5	5	5
>> Mitigation Scenario (Refores	2927	2927	2927	2927

Table 11-2, cont'd

Analysis of a Reforestation Mitigation Option, Country X

Item	1990	1992	1993	2030	
>>> STEP 5.1.1: STREAM OF COSTS ($/ha) OF REFORESTATION FOR A 1000 ha PLOT, 10-year ROTATION					
>> Initial Costs		800	500		
>> Recurrent (Maintenance etc.)		20	30	100	
>> Monitoring Costs		10	15	50	
>> Establishment Costs					
>> Total Costs		830	545	150	
>> Present Value of Cost	2927				
>>> STEP 5.2: BENEFITS ($/ha)					
>> Baseline Scenario (Wa	20	20	20	20	
>> Mitigation Scenario (	5663	5663	5663	5663	
>>> STEP 5.2.1: STREAM OF BENEFITS ($/ha) OF REFORESTATION FOR A 1000 ha PLOT, 10-year ROTATION					
>> Timber Product				100	
>> Non-timber benefits (fuel woo		10	15	50	
>> Non-timber benefits (raisin)		5.0	7.5	25.0	
>> Other benefits					
>> Total Benefits		15	22.5	175	
>> Present Value of Bene	5663				
>>> NET PRESENT VALUE OF	2736				
>>> STEP 6.1: TOTAL CARBON POOL (tC)					< Total >
>> Annual Incremental C Sequeste		101000	101000	101000	4040000
>> Baseline Scenario					
> Wasteland		3160000	3160000	3160000	
>> Mitigation Scenario		3362000	3463000	7200000	
> Wasteland		3002000	2923000	0	
> Reforested Land		360000	540000	7200000	
>>> STEP 6.2: TOTAL COSTS AND BENEFITS OF CSEQ ($)					<Present Value($)>
>> Incremental Net Benefit		2705883	2690883	2135883	25273370
>> Baseline Scenario Benefit		600000	600000	600000	5867430
> Cost		200000	200000	200000	1955810
> Benefit		800000	800000	800000	7823241
>> Mitigation Scenario Benefit		3305883	3290883	2735883	31140800
> Cost		3116617	3111617	2926617	30081684
> Benefit		6422500	6402500	5662500	61222484
>>> STEP 7: COST-EFFECTIVENESS INDICATORS					
>> Net Present Value of Benefits					
> $/tC					6.26
> $/ha.					632
>> Benefit of Reducing Atmospheric Carbon (BRAC)					
> $/tC-yr.					0.469
>> Initial Cost					
> $/tC					19.3
> $/ha.					1946
>> Endowment (Present Value of Costs)					
> $/tC					7.45
> $/ha.					752

APPENDIX 11-2
ESTIMATING NET PRESENT VALUE OF FORESTS MANAGED IN PERPETUAL ROTATION

This note explains the computation of the net present value (NPV) for a plantation or forest which is managed in perpetual rotations. We provide the formulas for computing the NPV for one rotation on a single plot, that for perpetual rotations on a single plot, and finally for a mosaic of perpetual rotations on multiple plots.

1. NPV per hectare for one rotation on one plot:

$$NPV = \sum_0^T (R_t - C_t)\, e^{-rt}$$

where R_t = Revenue per hectare in time t
C_t = Cost per hectare in time t
r = Rate of Discount
T = Rotation age in years
e = Natural log base

2. NPV per hectare for perpetual rotations on one plot (NPVP):

$$NPVP = NPV \,/\, (1 - e^{-rT})$$

Note that for coppice plantations, a rotation should be taken to mean the length of time until replanting. The coppice harvest and costs should be treated as intermediate output and costs.

3. NPV per hectare of perpetual rotations on multiple plots (NPVMP):

The NPV of perpetual rotations on multiple plots is

$$= NPVP \sum_0^T e^{-rT}$$
$$= NPVP\,(1-e^{-rT}) \,/\, (1 - e^{-r})$$

The NPVMP is obtained by dividing the above equation by T, which is

$$NPVMP = NPVP\,(1-e^{-rT}) \,/\, T\,(1 - e^{-r}).$$

APPENDIX 11-3
TRACE GAS EMISSIONS FROM FORESTRY

Some mitigation options in forestry reduce the emission of radiatively forcing trace gases such as CH_4, N_2O, NO_x, i.e., NO + NO_2, CO, and other NMHC. Such gases are emitted during (1) biomass-burning in forest clearing, (2) woodfuel combustion, (3) forest/savanna fires, (4) possibly when some termites digest biomass, (4) flooding of forest areas by dams, and (5) digestive processes of animals, mainly the ruminant group. However, the bulk of these emissions originate from biomass combustion.

Crutzen and Andreae (1990) estimated compound ratios of trace gas to total carbon released during biomass burning. For CH_4, CO, and NMHC, the ratio is to total carbon. For N_2O and NO_x, the ratio is for NO_x to C. The compound ratios for savanna burning are:

Ratios for trace gases:

Compound	Ratio
CH_4	0.007 - 0.013*
CO	0.075 - 0.125
N_2O	0.005 - 0.009
NO_x	0.094 - 0.148
NMHC	0.0131

* A more recent estimate by Delmas and Ahuja gives an estimate of 0.002 - 0.006, quoted in IPCC 1994.

Source: IPCC 1994.

C-CH_4 ratios for biomass fuels:

Fuel Type	C-CH_4/Total C Ratio	
Fuelwood	0.012	(0.009 - 0.015)
Agricultural Residues	0.005	(0.003 - 0.007)
Dung	0.017	
Charcoal combustion	0.005	(0.0014 - 0.0085)
Charcoal production	0.063	(0.040 - 0.090)

Source: Delmas and Ahuja, quoted in IPCC 1994.

To convert the ratios to full molecular weights, the emissions of CH_4 and CO are multiplied by 16/12 and 28/12 respectively, and the emissions of N_2O and NO_x are multiplied by 44/28 and 30/14 respectively.

Other trace gas parameters:

	Compound	Value	Units
Hydro dams	CH_4	0.157	MT/ha/yr**

** Based on the global average for lakes.

Source: Aselmann and Crutzen, 1990.

CHAPTER 12
AGRICULTURE

12.1 INTRODUCTION

This chapter presents methods for evaluating non-energy mitigation options for the agriculture sector. The primary sources of GHG emissions in this sector include animal husbandry, rice cultivation, fertilizer application, and soil carbon in cultivated soils. The latter are also a potential carbon sink.

Animal Husbandry. Animal husbandry results in methane emissions from two main sources: enteric fermentation (the digestive processes of animals) and manure management system facilities. Enteric fermentation emissions are driven principally by the quality and quantity of feed consumed by ruminant animals (cattle, buffalo, sheep, goats, and camels). Non-ruminant livestock (swine, horses, and poultry) produce a relatively small amount of methane from enteric fermentation.

Efforts to reduce methane emissions from enteric fermentation generally focus on options for improving production efficiency. Demand for animal products continues to increase globally, putting pressure on livestock production systems and competing land uses such as forestry. Improving production efficiency will not only help reduce these pressures and meet the growing demand but can also reduce methane emissions per unit of product produced. To be considered viable, emissions reduction strategies must also provide an economic return for the producer and be ecologically sustainable.

Manure-related emissions result from the anaerobic decay of organic material in the livestock manure. Manure management systems that promote anaerobic conditions produce the most methane: liquid/slurry storage facilities (tanks and pits) and anaerobic lagoons. While a relatively small percentage of livestock manure worldwide is managed in this manner, these systems are responsible for about 60% of global livestock manure methane emissions (USEPA, 1994). In contrast, management techniques which involve contact of the manure with air (e.g., uncollected on the range or spread directly on crops or pasture land) have limited methane production potential.

Efforts to reduce methane emissions from manure management facilities focus on options for recovering and using the methane produced. Techniques for reducing emissions should also maintain the fertilizer value of the material as well as provide a cost-effective means of handling the manure.

Rice Cultivation. Methane is emitted from flooded rice fields (both rain-fed and irrigated) due to the anaerobic decomposition of organic matter in the paddy soil. Current estimates of methane emissions from flooded rice fields worldwide range from 20 to 150 Tg per year (IPCC, 1992), making it one of the largest anthropogenic sources of methane emissions globally.[1]

The principal approach for reducing methane emissions from flooded rice cultivation is to modify growing practices. Research is ongoing to identify and evaluate options for reducing methane emissions by managing the input of organic material to the cropping system and/or modifying the flooding regimes to alter the anaerobic conditions that lead to methane formation.

While research on alternative growing practices appears promising, significant investigation remains to assess the complex set of factors that affect the methane emission process from flooded rice fields. In particular, the impacts of alternative growing practices on production as well as methane

---

[1] Most of the methane associated with rice cultivation is produced in irrigated and rain-fed wetland rice fields, which comprise over 75% of the area of global cultivated rice fields. Neither dry upland rice fields nor deepwater rice are believed to produce significant amounts of methane. Because 90% of worldwide rice production occurs in Asia (Braatz and Hogan, 1991), this region accounts for the majority of methane emissions from rice cultivation.

emissions over extended periods of time remain to be quantified. Because rice plays a critical role in the culture and diets of over one billion people globally, options for mitigating emissions must satisfy the following criteria: 1) improve rice productivity; 2) provide a positive economic return to the producer; 3) not require significant amounts of hard currency; and 4) be ecologically sustainable.

Fertilizer Application. It is well documented that N_2O is accumulating in the atmosphere, but the activities that are leading to this accumulation have yet to be well quantified. It has been suggested, however, that about 70% of N_2O emissions originate from soils (Bouwman, 1990 and IPCC, 1992). Consequently, perturbations to the soil nitrogen (N) cycle are believed to be contributing to the increase in atmospheric N_2O concentrations. While a variety of factors influence the N cycle in soils, the increase in N input to soil systems is one factor that has been quantified (Mosier, 1993).

Nitrogen can be added to soils for crop production through the application of N-based fertilizers and organic materials, by the cultivation of nitrogen-fixing plants, through the application of irrigation water containing dissolved N, and from precipitation. Experimental and monitoring data show that the application of N to cultivated soils increases emissions of N-based gases from the soil to the atmosphere. However, the transport and fate of N depends on numerous soil characteristics, the fertilizer formulation, fertility profile of the soil, moisture regime, and management. When applied to a crop, 100 kg of fertilizer N could result in 50 of the 100 kg of N harvested in the crop, and 50 lost by the combination of leaching (25 kg), surface runoff (5 kg), and gaseous losses (20 kg). A combination of conventional wisdom and limited research data indicates that 5% to 10% (1 to 2 kg) of the gaseous emissions will be N_2O. The remaining makeup of the gaseous emissions will include ammonia (NH_3), elemental N_2, and various oxides of N (NO_x).

Use of fertilizer N is essential for obtaining desired crop yields. Appropriate N management practices have the benefit of improving fertilizer N use efficiency, enhancing crop growth and quality, protecting surface water and groundwater quality, and reducing emissions of N_2O to the atmosphere. Some of the practices that should be included in a nitrogen management program are: 1) soil testing; 2) timing of N application to when it is needed by the crop; 3) accounting for the residual N in the soil and the N nitrification potential of the soil; 4) establishing a reasonable yield goal based on the soil/crop; and 5) accounting for N mineralization and N from legumes, animal manure and other organic waste, and irrigation water (CAST, 1992).

Although ongoing research clearly indicates that N-gas emissions can be reduced by implementing fertilizer management techniques, more information is needed to define and improve N fertilizer management practices to minimize losses of N_2O to the atmosphere. However, understanding of the processes involved has evolved well enough to permit the formulation of defensible mitigation strategies.

Soil Carbon in Cultivated Soils. Cultivated soils are believed to be a source of carbon emissions because cultivation of soils has been shown to lower their organic carbon content (Cole *et al.*, 1993; Duxbury and Mosier, 1993; Lee *et al.* 1993). The soil carbon loss occurs within about 25 years following the initial conversion of the native land cover to cultivated conditions. Following this initial loss, the amount of soil carbon will generally stabilize at a level consistent with the cultivation practice used.

Loss of soil organic carbon (SOC) occurs due to both enhanced mineralization of organic matter and erosion. The extent to which the loss of SOC contributes to increases in the atmospheric concentration of CO_2 is not known, however much (if not most) of the SOC is likely lost through erosion, by wind, or by water. For example, Lee *et al.* (1993) estimated that 35% of current SOC loss from soils in the U.S. cornbelt is by water erosion. Much of the eroded soil is deposited near the site of erosion, while some is transported to more distant locations with considerably different environmental conditions. Often, but not always, oxidation of SOC under these conditions is slower than at the original site. The uncertainty in the fate of SOC transported through erosion translates into an uncertainty in the estimate of the amount of CO_2 released from or sequestered by agricultural soils.

The use of alternative tilling practices can not only reduce the rate of SOC loss, it can also lead to a net accumulation of carbon in the soil. As such, changing tilling practices to accumulate SOC can help mitigate CO_2 emissions from other sources. The most promising opportunities for accumulating SOC are in areas where conventional tillage practices can be replaced with conservation tillage practices such as minimum till, no-till, much-till, etc.

12.2 POTENTIAL MITIGATION OPTIONS

12.2.1 Animal Husbandry

12.2.1.1 Enteric fermentation

The conditions under which livestock are managed vary greatly. Options for reducing emissions must be consistent with country-specific circumstances, including: animal management practices (including cultural traditions), feed resources, market conditions, and economic development priorities. One common strategy for reducing methane emissions is to increase animal production efficiency (e.g., milk production in cows, reproductive efficiency of cows maintained to produce calves, and the stamina and strength of draft animals). Virtually all efforts that improve animal productivity will reduce CH_4 emissions per unit of product produced.[2]

The ability to reduce methane emissions per unit of product has been demonstrated in various countries with intensive animal production systems. Experience has shown that proper veterinary care, sanitation, ventilation (for enclosed animals), nutrition, and animal comfort provide the foundation for improving production efficiency and reducing methane emissions. In many cases, focusing on these basics provides the best opportunity for improving production efficiency. Within this context, a variety of techniques can help improve animal productivity and reduce methane emissions per unit of product.

The following are the main strategies identified to date for reducing methane emissions from ruminant livestock. Specific approaches for each of these strategies are described in more detail in USEPA (1993a).

- **Improved Nutrition Through Mechanical and Chemical Feed Processing:** Improved nutrition reduces methane emissions per amount of product by enhancing animal performance, including weight gain, milk production, work production, and reproductive performance. Methane emissions per amount of digestible energy consumed by the animal may also be reduced. This option is applicable to accessible ruminant animals with limited feed resources. Assuming that feed digestibility is increased by 5%, methane emissions per unit product produced may decrease on the order of 10 to 25%, depending on animal management practices.

- **Improved Nutrition Through Strategic Supplementation and Other Methods**: Improved rumen function will reduce methane emissions per amount of feed consumed. Also, by providing additional microbial and/or by-pass protein to the animal, emissions per amount of product will be reduced by enhancing animal performance, including weight gain, milk production, work production, and reproductive performance. Improved rumen

---

[2] The methane emissions per unit of product produced vary substantially among regions and livestock production systems. For example, while a highly productive cow may emit 120 kg of methane per year, she may also produce 7,000 kg of milk, resulting in a release of 17 grams of methane per kilogram of milk produced. In comparison, a smaller cow on a straw-based diet may emit 50 to 60 grams of methane per kilogram of milk produced (Leng, 1991).

function may reduce methane emissions by about 5 to 10%. In addition, emissions per unit product may be reduced by 25 to 75% due to substantial increases in animal productivity (Leng, 1991).

- **Production Enhancing Agents:** Certain agents can act directly to improve productivity and reduce methane emissions per unit product. While various agents are under development, two agents that are currently available commercially are bovine somatotropin (bST) and anabolic steroid implants.

- **Improved Production Through Improved Genetic Characteristics:** Genetic characteristics are limiting factors mainly in intensive production systems. Continued improvements in genetic potential will increase productivity and thereby reduce methane emissions per unit product.

- **Improved Production Efficiency Through Improved Reproduction**: Large numbers of ruminant animals are maintained for the purpose of producing offspring. Methane emissions per unit product can be significantly reduced if reproductive efficiency is increased. Improved nutrition described above can improve reproduction. Additionally, there are options for addressing reproduction directly.

The technical applicability of these approaches to the main livestock management systems is summarized in Table 12-1. For example, for small-scale dairy and draft animal production systems and subsistence-mixed-farming production systems (generally found in developing countries), a range of options is applicable, including improved feed processing, improved nutrition through the use of supplements, and genetic improvements through breeding programs.

Additional criteria for assessing whether reducing emissions from ruminant livestock is a promising area include the size of the population of large ruminant livestock and the extent to which conditions support efforts to improve production efficiency. Economic development priorities that include the livestock sector are a useful complement to emission reduction efforts. Improved production and resulting incomes, for example, among small-scale dairy producers, can both reduce methane emissions per unit of product and improve rural economic conditions. Efforts to improve marketing arrangements and infrastructure are also a necessary complement to initiatives to reduce emissions. In addition, educational opportunities for livestock producers, e.g., on-farm from agriculture extension services, may be one component of efforts to promote the use of emissions-reducing practices.

In some countries, producer cooperatives have successfully created a climate that supports improvements in production efficiency. For example, cooperatives can provide training, assist in product marketing, and improve cash flow for producers, enabling them to invest in productivity-enhancing practices. Working with producer cooperatives to improve production in a manner that also reduces methane emissions may be an attractive approach for some countries.

12.2.1.2 Manure management system facilities

Manure management system facilities that promote anaerobic conditions produce the most methane, including liquid/slurry storage facilities (pits and tanks) and anaerobic lagoons. The preferred approach for reducing methane emissions from manure management facilities is to recover the methane produced and combust it; in the process the methane can be used as an energy source. Methane recovery technologies have been successfully used and demonstrated under a variety of conditions and have been shown to reduce emissions by up to 70 or 80% (USEPA, 1993a). Three main approaches have been identified. Table 12-2 summarizes the characteristics of these techniques.

Table 12-1. Applicability of Enteric Fermentation Emission Reduction Options to Animal Management Systems

	% of Animals (% of Emissions)	Feed Processing	Improved Nutrition	Production-Enhancing Agents	Genetic Improvement	Improved Reproduction
Intensive Dairy — Non-grazing	10-15% of animals (20-25% of emissions)	Processed feeds used routinely	Balanced rations used routinely	Candidate for additional implementation (e.g., bST)	Strong programs in place in main dairy countries	Strong programs in place in main dairy countries
Intensive Dairy — Grazing		(NA)	Candidate for targeted supplementation in selected areas with deficiencies	Candidate for additional implementation (e.g., bST)	Strong programs in place in main dairy countries	Strong programs in place in main dairy countries
Extensive Commercial Ranching	35-40% of animals (40-45% of emissions)	(NA)	Possible candidate for targeted supplementation in selected areas with deficiencies, depending on deliverability	Candidate for additional implementation	Candidate for additional implementation	(NA)
Non-Extensive Commercial Ranching		(NA)	Candidate for targeted supplementation in selected areas with deficiencies; candidate for defaunation	Currently used routinely	Candidate for additional implementation	Possible candidate for targeted implementation in cases with adequate access to animals
Feedlot Production	1-2% of animals (2-4% of emissions)	Processed feeds used routinely	Balanced rations used routinely	Currently used routinely where allowed	(NA)	(NA)
Small-scale Dairy and Draft	15-20% of animals (10-15% of emissions)	Candidate for additional implementation	Candidate for using molasses-urea blocks (MUB) and bypass protein feeds (BPF)	(NA)	Candidate for additional implementation	(NA)

Source: USEPA (1993b)

Table 12-2. Summary of Methane Recovery Techniques for Livestock Manure

Considerations	Covered Lagoons	Large-scale Digesters	Small-scale Digesters
Technique	Impermeable lagoon covers	Complete mix plug flow	Fixed dome Floating holder Flexible bag
Gas Quality	Medium quality (600-800 Btu/cf) (22-29 MJ/m^3)	Medium quality (600-800 Btu/cf) (22-29 MJ/m^3)	Medium quality (600-800 Btu/cf) (22-29 MJ/m^3)
Use options	Electricity generation boilers, refrigeration, other fertilizer, feed supplement, other	Electricity generation boilers, refrigeration, other fertilizer, feed supplement, other	Electricity generation domestic gas use fertilizer, feed supplement, other
Availability	Currently available	Currently available	Currently available
Capital requirements	Low/moderate	Moderate	Low
Technical complexity	Low technology	Moderate technology	Low technology
Applicability	Temperate, tropical flush systems; low %TS[2]	Temperate, tropical 2-15 %TS[2]	Temperate, tropical 7-15 %TS[2]
Methane reductions[1]	Up to 80%	Up to 70% or more	Up to 70%

1 These are reductions which may be achieved at an appropriate individual site.
2 Percent Total Solids (%TS) is a measure of the concentration of the manure in water.

Source: USEPA (1993b).

Covered Lagoons: Covered lagoons treat and store manure along with the large quantities of water used to wash the manure solids out of livestock housing facilities. The manure is treated under anaerobic conditions, resulting in the production of significant amounts of methane, which is recovered using an impermeable floating lagoon cover. The methane generated from these systems is often sufficient for the energy needs of large-scale, intensive farm operations. Because their technology and capital needs are relatively low, covered lagoons may be appropriate for large farm operations in developing countries, especially those which need the high-quality liquid fertilizer that lagoons produce. The use of lagoons in arid regions, however, may be constrained by their high water requirements.

Small-scale Digesters: Digesters are designed to enhance the anaerobic decomposition of organic material and to optimize methane production and recovery. Small-scale digesters typically require a small amount of manure and are relatively simple to build and operate. As such, they are an appropriate strategy for small- to medium-confined or semi-confined farm operations. These digesters are also well-suited for regions with technical, capital, and material resource constraints, and have already been implemented successfully in countries such as China and India (United Nations, 1984). The digesters offer additional benefits in agricultural regions where manure is currently burned for fuel, as the generated methane is a cleaner and more efficient fuel. In addition, digested manure retains most of its fertilizer value. Small-scale digesters generally operate best in temperate and tropical areas.

Larger-scale Digesters: These digesters are also designed to enhance anaerobic decomposition and optimize methane recovery but have larger capacities and are often more technologically advanced. They are generally heated and can operate in relatively cold regions. Because larger-scale digesters require greater capital investment, are more complex to build and operate, and require large concentrations of manure, they are best suited for large livestock operations. These technologies are especially suitable at operations which handle manure as liquids (less than 10% solids) or slurry (10 to 20% solids).

The emissions reductions that can be achieved are a function of the amount of manure currently handled in a manner that produces methane emissions. The options will be most effective when conditions in the country support efforts to recover the methane and use it as an energy source. The most promising opportunity for implementing methane recovery is a large confined animal production facility using a liquid manure handling system. High local energy prices (which would increase the value of the methane recovered) would make these options attractive. Small-scale implementation can also be attractive, particularly if an alternative energy source is desired (e.g., due to depletion of forest resources).

In addition to recovering methane from manure facilities, methane emissions can be reduced by keeping the manure under aerobic conditions. Manure spreading on soils and composting are the primary approaches for maintaining aerobic conditions. When undertaking spreading, care must be exercised in the rate of application on land to protect groundwater quality and to prevent unwanted runoff. Also, spreading must generally be restricted during periods when the ground is frozen, making it unfeasible in some areas.

Composting is a good alternative for managing manure because the compost is a valuable fertilizer. Care must be taken to prevent unwanted runoff from the compost facility. Additionally, the potential for methane to be emitted from unwanted anaerobic conditions that may develop within the compost remains to be assessed.

12.2.2 Rice Cultivation

Efforts to date have identified a number of approaches which could reduce methane emissions from rice cultivation while maintaining the productivity of the rice fields (IPCC, 1990; Braatz and Hogan,

1991). Two of the most promising approaches under investigation are changes in nutrient management and water management practices (USEPA, 1993b):

- **Nutrient Management.** Research indicates that using N fertilizers and reducing the use of raw organic materials as fertilizer can reduce methane emissions from rice fields. Additionally, the form of the N fertilizer and the method of application may be important. This option may be promising because N fertilizers are already a major nutrient source for flooded rice fields in Asia. The major constraints to this option are the cost of N fertilizers, which may be prohibitive in some regions, and the existence of traditional fertilization techniques, which may be difficult to change. Additionally, care should be taken to apply N fertilizers in a way that N_2O emissions are minimized (see below).

- **Water Management.** Intermittent draining of rice fields during the growing season or between croppings appears to decrease methane production, as does increasing the water percolation rate in the fields. These methods may be technically feasible in lowland regions and flatland irrigated areas, which have secure and controllable water supplies. Proposed changes in water management practices must be researched carefully in order to avoid decreasing productivity.

The impacts of these approaches on methane emissions and rice production remain to be quantified fully. Research indicates, for example, that there are interactions among fertilizer type, application method, and soil type that may affect methane emissions. Additionally, the impacts of intermittent field drainage on weed growth and rice production are still being researched. Examples of recent and ongoing research include Kern *et al.* (1995), Olszyk *et al.* (1993); Masscheleyn *et al.* (1993); and Wang *et al.* (1993a and 1993b).

Efforts are also ongoing to examine other avenues for reducing emissions, such as developing rice cultivars which result in lower methane emissions, and altering existing rice cultivation practices, such as tillage and seeding techniques (Braatz and Hogan, 1991). Research in this area is continuing at the International Rice Research Institute and among key rice-growing nations. Countries are encouraged to evaluate options they believe promising as part of their mitigation analyses.

12.2.3 Fertilizer Application

Although the factors that affect N_2O emissions from N application are not completely understood, there are several nitrogen management techniques known to be effective in reducing overall N-based emissions. Control of the nitrate form of N, NO_3, is the critical aspect of managing N_2O emissions and is the central focus of the management techniques listed below:

- Test soils to determine N fertilizer needs.
- Establish yield goals based on site and crop characteristics.
- Set fertilizer rate and timing of application.
- Select N fertilizer formulation related to yield, leaching and runoff potential, and emissions.
- Account for animal manure and other organic waste where used in the fertility program.
- Use winter cover crops for removal of residual N.
- Place N deeper in the soil.
- Implement irrigation water management techniques.
- Use nitrifying and denitrifying inhibitors.
- Test soil for pH (and lime).

12.2.4 Soil Carbon in Cultivated Soils

While there are uncertainties in the estimates of CO_2 emissions from cultivated soils, the use of conservation tillage techniques has been shown to be effective in reducing SOC loss and, in some cases, can lead to SOC accumulation (Kern and Johnson, 1993). As alternatives to conventional tillage practices, conservation tillage techniques have been developed to reduce water and wind erosion, conserve soil moisture, and reduce fuel requirements. While they have the potential to reduce production costs, conservation tillage techniques require a higher level of farming skill.

Conservation tillage systems manage crop residues with a reduced amount of tillage and, in some cases, no tillage. In a mulch till approach, a mulch of crop residues is maintained to protect the soil against raindrop impact and wind, slow evaporation, increase water storage, and slow organic matter decomposition. No-till approaches consist of no tilling between harvest to planting time. No-till can be combined with the addition of a cover crop to further protect the soil. Table 12-3 summarizes the practices for conventional tillage and three examples of conservation tillage systems for continuous corn production. Although the examples were developed for U.S. conditions, the concepts are applicable for all cultivation practices.

The costs and benefits of these mitigation approaches remain to be assessed. Research in this area is continuing and countries are encouraged to evaluate options they believe promising as part of their mitigation analyses.

12.3 CONSTRUCTING BASELINE SCENARIOS

The IPCC/OECD emissions inventory methodologies identify the inputs needed to estimate GHG emisions for 1990 or an alternative base year (IPCC/OECD, 1994). Baseline scenarios of future emissions may be developed by forecasting the key variables that drive the emissions in each area analyzed. Ranges of assumptions should be developed for those parameters or estimates that are uncertain, and the implications of the uncertainty should be considered.

12.3.1 Animal Husbandry

The baseline scenario of future emissions can be developed using a simple method or a complex method. The simple method is driven by scenarios of the future production of animal products, such as milk, meat, and draft power. For each animal type, an average emissions factor per unit of production is estimated for 1990 based on the 1990 inventory. The emissions factors are then multiplied by estimated future production to estimate future emissions as follows:

$$\text{Emissions}_{i,t} = \text{Production}_{i,t} * \text{Emissions Factor}_{i,1990}$$

where:

$\text{Emissions}_{i,t}$ is the estimate of emissions for animal type i in year t;
$\text{Production}_{i,t}$ is the estimate of production for animal type i in year t; and
$\text{Emissions Factor}_{i,1990}$ is the estimate of the emissions factor per unit of production for animal type i in 1990.[3]

[3] The emissions factors discussed here are emissions factors per unit of product produced, for example, kg of methane emitted per ton of milk produced.

Table 12-3. Operations for Conventional and Conservation Tillage: U.S. Continuous Corn Production

Operation	Conventional Tillage	Conservation Tillage		
		Mulch Till	No-till	No-till & Cover
Tandem Disk	Y			
N Fertilizer	Y	Y	Y	Y
Tandem Disk	Y	Y		
Field Cultivator	Y	Y		
Row Planter	Y	Y	Y	Y
Rotary Hoe	Y			
Row Cultivator	Y	Y		
Row Cultivator	Y			
Harvest	Y	Y	Y	Y
Tandem Disk	Y			
P Fertilizer	Y	Y	Y	Y
Twisted Point Chisel	Y	Y		
Plant Winter Wheat				Y
Shred, Kill Wheat				Y

Source: Lee *et al.*, 1993.

This simple approach assumes that the emissions factor is not changing over time. The principal input required is the scenario of future production.

A more complex analysis should be conducted if the emissions factor per unit of production is expected to change in the future. A detailed assessment of the livestock production system is required, including an explicit assessment of how changes in production efficiency (e.g., increases in milk production per cow) will affect the size and performance of the overall national herd. Trends in the production of animal products, production levels per animal, and manure management practices should be considered.

12.3.2 Rice Cultivation

Future methane emissions can be estimated using forecasts of the growing activity. Future changes in amount of area under cultivation and water management conditions must be considered. If locally-derived emissions factors are used that reflect differences in conditions such as fertilization practices, the baseline changes in these conditions should also be forecast. Data for these forecasts should be drawn from national agriculture planning studies.

12.3.3 Fertilizer Application

As discussed above, the information needed to accurately estimate baseline N_2O emissions is not available. The IPCC/OECD emissions inventory guidelines allow a wide range of baseline emissions estimates to be developed using the following approach:

$$N_2O\ Emissions_{Low\ to\ High} = N\ Applied * Emissions\ Coefficient * 44/28$$

where:

N Applied = the tons of N applied to soils from chemical fertilizers, organic sources (animal manure and crop residues), and from biological sources (leguminous crops);

Emissions Coefficient ranges from a Low of 0.0005 to a High of 0.039, with a Median estimate of 0.0036; and

44/28 is the ratio of the molecular weight of N_2O to the molecular weight of N in N_2O.

The very large range of nearly two magnitudes between the low and high coefficients demonstrates the uncertainty in the current information available for making the emissions estimates.

Sources of data for estimating the N applied are listed in IPCC/OECD (1994). Forecasts of N applied would be needed to develop a baseline of future emissions.

12.3.4 Soil Carbon in Cultivated Soils

The IPCC/OECD emissions inventory guidelines provide a method for estimating CO_2 emissions from SOC in cultivated soils. To evaluate the mitigation potential of switching to conservation tillage practices, a baseline scenario of SOC characteristics and tillage practices is needed.

The potential to accumulate SOC depends on highly site-specific climate, soil, and crop management factors. Simply stated, the change in the SOC content over time is a function of the carbon input rates and carbon turnover rates for each soil layer, as follows (Cole *et al.*, 1993).

The decomposition rate depends on climate, soil texture, tillage practices, crop residue characteristics, and other factors. For sites with soil erosion by wind or water, it also inherently includes loss of SOC through transport of soil off-site. In this case, it does not necessarily represent the rate at which SOC is oxidized in CO_2. The uncertainty in the rate at which C is released or sequestered introduced by not knowing the fate of the SOC in eroded soil can be quantified by using an erosion model to estimate the amount of SOC removed by erosion, and then assuming that all or none of this SOC is released to the atmosphere.

12.4 ANALYSIS OF MITIGATION OPTIONS

12.4.1 Enteric Fermentation

- **Identify Target Subgroups to be the Focus of the Emissions Reduction Effort and Refine the Emissions Estimates for the Subgroup**

  The purpose of this step is to identify subgroups of the livestock population that are best to target for emissions reductions. Experts in animal nutrition and production within the country must be involved in this assessment. The following is recommended to identify subgroups of the national cattle and buffalo populations that are most promising for emissions reductions:

  - identify cattle and buffalo populations with relatively low levels of production efficiency;

  - identify under-utilized or inefficient use of conventional and non-conventional feed resources that could be used for the population with low levels of production efficiency;

  - identify the segment of producers with low production efficiency animals that can adopt improved production practices (e.g., those that are in organized cooperatives); and

  - identify the segment of these producers that are a priority for economic development or for improvements in markets or marketing approaches, which would take advantage of the increased production efficiency of these animals.

  For the subgroups that are promising candidates for emissions reductions, detailed data should be collected for purposes of refining the methane emissions estimates. The Tier 2 (detailed) emissions estimating procedure described in the IPCC/OECD inventory method (or a comparable method) should be used. In addition to livestock population data, the following information is required for the detailed assessment:

  - average weight of each animal type (kg);

  - average weight gain per day (kg);

  - feeding situation: confined animals, animals grazing good quality pasture, and animals grazing over very large areas;

  - milk production per day (kg/day);

  - average amount of work performed per day (hours/day);

  - percent of mature female cows that give birth in a year; and

  - feed digestibility (%).[4]

[4] Feed digestibility is defined as the proportion of energy in the feed that is not excreted in the feces. Common ranges for feed digestibility for cattle are 50% to 60% for crop by-products and rangelands; 60% to 70% for good pastures, good preserved forages, and grain-supplemented forage-based diets; and 75% to 85% for grain-based diets fed in feedlots.

These data should be obtained from national livestock experts. A detailed baseline of future emissions should be developed based on estimates of changes in these characteristics as well as future production levels.

- **Evaluate the Mitigation Options for the Subgroup**

  The impact of each of the mitigation options applicable to each target subgroup is estimated as follows:

  - Identify how the mitigation option will affect the livestock characteristics used to estimate emissions. For example, the feed digestibility may be increased by 5%, or the milk production per cow may be increased by 15%.

  - Estimate new emissions factors per animal using the new livestock characteristics (the same method used to develop the baseline emissions estimates should be used). Estimate a new age structure of the livestock population if the mitigation option affects growth rates, reproduction, or mortality. Using these emissions factors per animal and the new age structure of the livestock population, estimate new emissions factors per unit of product produced.

  - Estimate how the mitigation option will affect the overall production of animal products. In some cases, overall production efficiency may increase as the result of improved production efficiency. Overall production will generally be constrained by feed availability and cost, however.

  - Estimate the change in emissions using both the estimated change in the emissions per unit of production and the estimated change in production.

  The costs of the options must be estimated in terms of training costs, increased feed processing costs, infrastructure investment costs, and other costs. Estimates of costs will depend on local conditions and the details of the projects and programs being considered. These costs will be offset by benefits in terms of increased production efficiency and improved standards of living for rural populations. The costs and benefits should be estimated from the perspective of the individual producer as well as nationally. Only those options that are made to be profitable from the perspective of the individual producer will likely be adopted. ATI (1992) is an example of the type of assessment required in this step.

12.4.2 Manure Management System Facilities

- **Identify Target Subgroups to be the Focus of the Emissions Reduction Effort and Refine the Emissions Estimates for the Subgroup**

  Experts in animal production and manure management within the country must be involved in this assessment. The following is recommended to identify subgroups of livestock that are most promising for emissions reductions:

  - identify livestock populations managed in large production facilities (typically large dairies and hog production facilities) that manage manure in liquid systems;

- identify livestock populations (typically small-scale dairy and draft cattle) whose manure is burned for fuel; and

- identify the segment of the large and small producers that would benefit from having an additional energy source, i.e., the methane recovered from the manure.

For the subgroups that are promising candidates for emissions reductions, detailed data should be collected for purposes of refining the methane emissions estimates. The Tier 2 (detailed) emissions estimating procedure described in the IPCC/OECD inventory method (or a comparable method) should be used. The following information is required to conduct this assessment:

- annual average population (number of head) and climate region (cool, temperate, and warm);

- average daily manure volatile solids (VS) excretion (kg per day of dry matter);[5]

- methane producing potential (B_o) of the manure (m^3 of methane per kg of VS);

- manure management system facility usage (percentage of manure managed with each manure management system facility); and

- methane conversion factors (MCFs) for each manure management system used (the MCF defines the portion of B_o that is achieved).

A detailed baseline of future emissions should be developed based on estimates of future production levels and manure management practices. The approach recommended for developing a scenario of future emissions is the same as discussed above for enteric fermentation.

- **Evaluate the Mitigation Options for the Subgroup**

  The impact of each of the mitigation options on emissions can be estimated as the amount of methane that will be recovered that would otherwise have been emitted. USEPA (1993c) presents an example of how this analysis can be performed.

  The costs of the options must be estimated in terms of training costs, increased labor costs, infrastructure investment costs, and other costs. Estimates of costs will depend on local conditions and the details of the projects and programs being considered. These costs will be offset by benefits in terms of energy produced and the maintenance of the fertilizer value of the manure. Additional benefits may include reduced pressure on forest resources in rural areas for fuel and improved standards of living for rural populations. The costs and benefits should be estimated from the perspective of the individual producer as well as nationally.

---

[5] Volatile solids (VS) is a measure of the degradable organic material in livestock manure. The VS excretion rate is driven by feed intake and feed digestibility. Feed intake may be estimated using the Tier 2 emissions inventory method for enteric fermentation (IPCC/OECD, 1994).

12.4.3 Rice Cultivation

Methods for evaluating mitigation options for methane emissions from rice cultivation are in the early stages of development. As discussed above, research is ongoing to identify and evaluate various approaches for reducing emissions. Once reliable information is available on various options, the following general approach may be used to evaluate the options.

- **Identify Target Subgroups to be the Focus of the Emissions Reduction Effort and Refine the Emissions Estimates for the Subgroup**

  To identify the most promising areas for emissions mitigation, a list of candidate mitigation options is needed, and the applicable conditions for each option must be identified and compared with a database of growing conditions in the country. For example, modifying water management to periodically drain the fields may only be practical in areas with secure irrigation supplies. A detailed database on national growing conditions, including irrigation status, would be needed to identify those areas that met the applicability criteria for the option. Such as database could be organized in a Geographic Information System (GIS), which would also facilitate the evaluation.

  Once the target areas are identified, the emissions baseline for each should be re-examined. If possible, detailed site-specific measurement data from several years for these specific areas should be used to develop emissions factors. Additionally, estimates of days flooded should be developed for these specific target areas. Using these data, a detailed baseline of emissions for the target areas should be developed.

- **Evaluate the Mitigation Options for the Subgroup**

  Data on the impacts of each mitigation option on the emissions factors should be used to estimate the changes in the emissions. Additionally, the costs of the options, including the impacts on rice production, must be assessed. The data needed to evaluate impacts on emissions and costs are not yet available, however. Recent analyses based on ongoing research results that demonstrate this approach are presented in Kern *et al.* (1995), Bachelet and Neue (1993), and Bachelet, *et al.*, (1993).

12.4.4 Fertilizer Application

Evaluation of mitigation options for managing N_2O emissions should begin with the selection of land unit subdivisions to be analyzed. The analysis needs to encompass all of the major land resource (physiographic) regions in which agricultural crops are grown. These regions should be established on the basis of groups of similar physical, cultural, structural, and meteorological features. The features involved in establishing physiographic regions include the following:

- physical: soils (soil nitrate levels, bulk density, texture classes, etc.), slope;
- cultural: crops (type, yield), tillage, nitrogen application (all sources) and crop nitrogen uptake, conservation practices;
- structural: irrigation systems, land forming; and
- meteorological: precipitation, temperature.

Each of the above features will require specific data. Additionally, the analyst will need to determine the size of the areas according to the extent of these features and their overall importance to potential mitigation options.

Once the physiographic regions are established, it is necessary to select the combinations of features that represent the majority of existing agricultural production systems within each region and organize them into categories of typical production systems, such as non-irrigated conventional till maize (plant date & harvest date) grown on loamy sand soils (profile description needed) having an average land slope of 2%, average annual precipitation of 600 mm (seasonal and daily time steps will be needed), with a particular nitrogen management program (types, rates, timing, placement).

Through the use of the analytical technique formulated in the computer model NLEAP ("Nitrate Leaching and Economic Analysis Package"), the gross emission of N gasses to the atmosphere can be determined (Shaffer *et al.*, 1991). NLEAP accounts for movement of N throughout the primary flow circuit of the nitrogen cycle including the amount of N-gases emitted to the atmosphere. The specific amount of N_2O in the emitted gas, under a range of conditions, is not known with certainty and is the subject of further research. However, the total amount of gas emitted can be calculated through use of NLEAP. With additional development of NLEAP along with collection of experimental and field data, the amount of N_2O in the total amount of gas emitted can be estimated based on the site variables.

Two (or more) runs of the NLEAP model could constitute the analytical process for evaluating mitigation options. The first model run would simulate the existing agricultural production systems, and the second would incorporate the options that contain changes in the way N is managed. The change in N-based gas emissions can be determined. This process is repeated for all physiographic regions and all typical agricultural production systems within the regions. Results from this process are summed to obtain totals for the country. The elementary cost of adjusting nitrogen management techniques can also be determined through use of NLEAP.

Although nitrogen management for N_2O reduction is the focus of these evaluations, it is essential that the overall nutrient requirements of crops be evaluated before making decisions about how to best manage N. Without this approach, it is possible to reduce N-gas release into the atmosphere while increasing the movement of nitrates in the ground or surface water, thus creating the undesirable tradeoff. Also, nitrogen management should be balanced within the total fertility program needed for individual crops. Phosphorus, for example, could become overloaded in the soil profile if care is not excised. Implementation of comprehensive nutrient management programs will provide guidance needed to avoid problems. Nitrogen management options need to be integrated with the other mitigation options to identify optimal strategies. Finally, mitigation options need to be further integrated into an overall ecologically-based plan.

The costs and benefits should be estimated from the perspective of the individual producer as well as nationally since only those options that are made to be profitable from the perspective of the producer will likely be adopted.

12.4.5 Soil Carbon in Cultivated Soils

Models such as EPIC and CENTURY may be used to evaluate the impacts of conservation tillage on SOC[6]. These models require detailed data on carbon inputs and factors that affect the turnover rate of SOC, including climate, soil texture, tillage practices, and crop residue characteristics. These data can be developed from national data bases when available or from site-specific studies. Forecasts of future

---

[6] The EPIC model is described in Sharply and Williams (1990a and 1990b). The CENTURY model is described in Parton *et al.* (1992).

practices should consider ongoing trends in the use of various tillage techniques. Additionally, because weather conditions are an important factor affecting SOC, scenarios of future changes in climate (temperature and precipitation) may also be considered in the assessment (see, e.g., Woomer, 1993).

First, the most promising areas for switching to conservation tillage should be defined as the target sub-areas. Data for estimating the current and potential future SOC levels in the target sub-areas should be collected. This step may require field data collection.

Using the data collected, the SOC model(s) chosen should be calibrated to reproduce the current SOC characteristics of the areas based on past and current management practices. Once the model is calibrated to estimate SOC for past and current management practices, model parameters should be estimated using field data on the impacts of conservation tillage techniques on SOC. If possible, the results of local field studies of conservation tillage techniques should be used. If necessary, long-term field research should be initiated to develop the parameters needed to specify the models for the relevant local conditions. Examples of recent analyses of this type include Woomer (1993), Cole *et al.* (1993), Lee *et al.* (1993), and Kern and Johnson (1993).

Generally, the detailed analysis must be performed on a representative set of selected sites. The results for these sites are then extrapolated to the target subgroups.

12.5 CONSTRUCTING MITIGATION SCENARIOS

The purpose of constructing mitigation scenarios is to estimate future emissions with implementation of the selected mitigation options. It is recommended that these scenarios be developed in detail for the target subgroups for specific years, such as 2000 and 2010. The same models or techniques should be used to estimate both the baseline and the mitigation emissions scenarios for each of the sources.

Two separate mitigation scenarios may be developed: (1) the technical feasibility scenario; and (2) the likely achievable scenario. The technical feasibility scenario provides an upper bound on the mitigation potential of the options examined. The likely achievable scenario takes into account the extent to which the mitigation options could realistically be adopted or implemented over time. The analyst should consider a range of penetration rates for the techniques based on penetration rates of similar technologies under comparable circumstances. A study by ATI (1992) for reducing methane emissions from ruminant livestock provides an example of how adoption rates can be included in an analysis. The analyses should also be explicit regarding the policies that might be undertaken to promote the adoption of the techniques and how the policies may affect the penetration rate.

12.6 MITIGATION POLICIES

12.6.1 Approaches for Implementing Mitigation Options

There are several key approaches for promoting the implementation of mitigation options in the agriculture sector. Generally, these approaches help to overcome the main barriers to the implementation of the mitigation options, which include lack of information; lack of training; and inadequate economic incentives or economic resources (such as capital). Each approach is summarized briefly.

Demonstration Projects. Demonstration projects can be used to show that recommended technology practices are cost-effective for producers. Research initiatives may be combined with demonstration projects to identify the preferred practice(s) for reducing emissions in a given location. Such projects would need to be targeted to the subgroups discussed above and may be needed in several regions with differing conditions. Demonstration projects are usually needed because agriculture producers are necessarily risk-averse and, as a consequence, are slow to adopt changes in production practices.

Pilot Projects. Pilot projects are a step beyond demonstration projects. As such, pilot projects provide an opportunity to show how the proposed emissions reduction technique can operate on a larger scale. Necessary experience in project management, regional coordination, benefits assessments, and logistics can be obtained from pilot projects prior to implementing programs nationally.

Education. Education programs will be needed to disseminate information on the emissions reduction techniques. Education may be of two types:

- Formal education through the public school system and universities would target producers, agriculture professionals, and policy-makers.
- Extension services include training and assistance from extension personnel. Services may be provided on-farm or off-farm, at research or demonstration sites.

Incentives. Incentives may be needed to promote the use of the techniques. Examples include direct financial assistance for expenses (e.g., capital costs or materials costs); subsidies for key inputs (e.g., for equipment); credit on favorable terms for investments in emissions reductions; and recognition for emissions reductions achieved.

Direct Provision. The inputs needed to reduce emissions may be provided directly to producers at reduced cost or for free. Examples include providing information on the effectiveness of various techniques; providing training; providing infrastructure needed for undertaking mitigation options (e.g., irrigation for rice fields). In some cases, the program may involve commercializing products through the private sector so that the government's involvement and subsidies are eliminated over time.

In addition to these approaches, various mitigation techniques could be required. For example, methane recovery from large manure treatment lagoons or conservation tillage on erodible land could be required. Requiring changes in production practices among small-scale rural producers is not likely to be preferred, however.

12.6.2 Considerations in Developing a National Strategy

Key considerations in developing a national strategy for mitigating emissions from agriculture include the following:

Scope. The scope of the emissions reduction strategy should be well defined, including the subgroups to be targeted; the specific options to be promoted; the time frame over which the options will be encouraged; and the approaches to be used to promote the use of the emissions reduction options. Realistic penetration rates of the emissions reduction options should be used in the assessment of the total emissions reduction likely to be achieved. Experience with previous programs implemented in the relevant sectors in the country could be used to assess likely adoption rates.

Coordination with Other Development Objectives. The goal of reducing emissions from agriculture can often be combined with rural economic development objectives. The most promising options for reducing emissions will be those that also lead to improvements in the standard of living of rural populations. The national objectives for rural economic development and assistance must be considered during the development of a strategy for reducing these emissions.

Food Security. Any efforts to influence the agriculture sector must consider impacts on food security and trade balances. Options that reduce rice production are not likely to be economically feasible for most major rice-growing nations. Alternatively, improvements in livestock production efficiency could enhance food security and reduce reliance on imports. Care must be taken to ensure that locally-available resources are used to improve food security; options that would result in over-reliance on imports may be inadvisable.

REFERENCES

ATI (Appropriate Technology International) (1992), *Reducing Methane Emissions from Ruminant Livestock: India Prefeasibility Study,* prepared for the Global Change Division of the U.S. Environmental Protection Agency, Washington, D.C.

Bachelet, D. and H.U. Neue (1993), "Methane Emissions from Wetland Rice Areas of Asia," *Chemosphere*, 26:219-237.

Bachelet, D., J.S. Kern, and M. Tolg (1993), "Modeling the rice carbon budget in China using GIS Technology," in Huang, X. *et al.* (eds.), *Proceedings of the 3rd International Workshop on Geographic Information Systems,* Chinese Academy of Sciences, Beijing, China, 3:31-53.

Blayney, D.P. and R.F. Fallert (1990), *Biotechnology and Agriculture: Emergence of Bovine Somatotropin (bST),* U.S. Dept. of Agriculture, Washington, D.C., Staff Report AGES 9037.

Braatz, Barbara V. and Kathleen B. Hogan, eds. (1991), *Sustainable Rice Productivity and Methane Reduction Research Plan*, U.S. Environmental Protection Agency/Office of Air and Radiation, Washington, D.C.

CAST (U.S. Council for Agricultural Science and Technology), 1992. "Preparing U.S. Agriculture for Global Climate Change". Report 119.

Cole, C.V., K. Paustian, E.T. Elliott, A.K. Metherell, D.S. Ojima, and W.J. Parton (1993), "Analysis of Agroecosystem Carbon Pools," in J. Wisniewski and R.N. Sampson (eds.) *Terrestrial Biospheric Carbon Fluxes: Quantification of Sinks and Sources of* CO_2, Kluwer Academic Publishers, Netherlands.

Ensminger, M.E. (1983), *The Stockman's Handbook*, The Interstate Printers & Publishers, Danville, Illinois.

FAO (Food and Agriculture Organization) (various years), *FAO Production Yearbook*, FAO, United Nations, Rome, Italy.

IPCC (1990), Methane Emissions and Opportunities for Control: Workshop Results of Intergovernmental Panel on Climate Change, Response Strategies Working Group, coordinated by USEPA/OAR and Japan Environment Agency, September, 1990.

IPCC (1992), *Science Assessment*, Intergovernmental Panel on Climate Change Work Group I, 1992.

IPCC/OECD (Intergovernmental Panel on Climate Change/Organization for Economic Cooperation and Development) (1994), *Greenhouse Gas Inventory Reference Manual: IPCC Draft Guidelines for National Inventories*, OECD, Paris, France.

Kern, J.S., and M.G. Johnson (1993), "Conservation Tillage Impacts on National Soil and Atmospheric Carbon Levels," *Soil Science Society of America Journal*, Vol 57, pp. 200-210.

Kern, J.S., D. Bachelet, and M. Tolg (1995), "Organic Matter Inputs and Methane Emissions from Soils in Major Rice Growing Regions in China," in *Advances in Soil Chemistry*, Lewis Publishers.

Lee, J.J., D.L. Phillips, and R. Liu (1993), "The Effect of Trends in Tillage Practices on Erosion and Carbon Content of Soils in the US Corn Belt," in J. Wisniewski and R.N. Sampson (eds.) *Terrestrial Biospheric Carbon Fluxes: Quantification of Sinks and Sources of CO_2*, Kluwer Academic Publishers, Netherlands.

Leng, R.A. (1991), *Improving Ruminant Production and Reducing Methane Emissions from Ruminants by Strategic Supplementation*, prepared for the US Environmental Protection Agency, Washington, D.C., May, 1991. EPA/400/1-91/004.

Masscheleyn, P.H., R.D. Delaune and W.H. Patrick, Jr. (1993), "Methane and nitrous oxide emissions from a flooded rice soil: Effect of soil oxidation-reduction status," *Chemosphere*, 26:251-260.

Mosier, A.R. (1993), "Nitrous Oxide Emissions from Agricultural Soils," in A.R. van Amstel (ed.) *Methane and Nitrous Oxide*, National Institute of Public Health and Environmental Protection, Bilthoven, The Netherlands, pp. 273-285.

Olszyk, D. *et al.* (1993), *Effects of UV-B and Global Climate Change in Rice: Second Annual Program Progress Report*, U.S. Environmental Protection Agency, Environmental Research Laboratory, Corvallis, Oregon, EPA/600/R-93/128.

Owen E., and M.C.N Jayasuriya (1989), "Recent Developments on Chemical Treatment of Roughages and their Relevance to Animal Production in Developing Countries," in *Feeding Strategies for Improving Productivity of Ruminant Livestock in Developing Countries*, International Atomic Energy Agency, Vienna, Austria, pp. 205-230.

Parton, W.J. *et al.* (1992), *CENTURY Users Manual*, Colorado State University, Fort Collins, Colorado.

Shaffer, M.J., A.D. Halvorson, and F.J. Pierce, 1991, "Nitrate Leaching and Economic Analysis Package (NLEAP): Model description and application. In: R.F. Follett, D.R. Keeney, and R. M. Cruse (eds.). Managing Nitrogen for Groundwater Quality and Farm Profitability. pp 285-322. Soil Science Society of America, Inc., Madison, WI.

Sharpley, A.N. and J.R. Williams (eds.)(1990a), *EPIC--Erosion/Productivity Impact Calculator: 1. Model Documentation,* U.S. Department of Agriculture Technical Bulletin No. 1768, Washington, D.C.

Sharpley, A.N. and J.R. Williams (eds.)(1990b), *EPIC--Erosion/Productivity Impact Calculator: 2. User Manual,* U.S. Department of Agriculture Technical Bulletin No. 1768, Washington, D.C.

United Nations (1984), *Updated Guidebook on Biogas Development*, Energy Resources Development Series No. 27 New York, Economic and Social Commission for Asia and the Pacific, Thailand.

USDA (U.S. Department of Agriculture) (1987), *Economic Impacts of the European Economic Community's Ban on Anabolic Implants,* Appendices A-D, Food Safety and Inspection Service, October 1987.

USEPA (1993a), *Options for Reducing Methane Emissions Internationally. Volume I: Technological Options for Reducing Methane Emissions*, U.S. Environmental Protection Agency, Office of Air and Radiation, K. Hogan (ed.), Washington, D.C.

USEPA (1993b), *Options for Reducing Methane Emissions Internationally. Volume II: International Opportunities for Reducing Methane Emissions*, U.S. Environmental Protection Agency, Office of Air and Radiation, K. Hogan (ed.), Washington, D.C.

USEPA (1993c), *Opportunities to Reduce Anthropogenic Methane Emissions in the U.S.*, U.S. Environmental Protection Agency, Office of Air and Radiation, K. Hogan (ed.), Washington, D.C.

USEPA (1994), *International Anthropogenic Methane Emissions: Estimates for 1990*, Report to Congress, U.S. Environmental Protection Agency, Office of Policy, Planning and Evaluation, M.J. Adler (ed.), Washington, D.C.

Wang, Z.P., R.D. Delaune, P.H. Masscheleyn, and W.H. Patrick, Jr. (1993a), "Soil redox and pH effects on methane production in a flooded rice soil," *Journal of Soil Science Society of America*, 57:382-385.

Wang, Z.P., R.D. Delaune, C.W. Lindau, and W.H. Patrick, Jr. (1993b), "Methane production from anaerobic soil amended with rice straw and nitrogen *Fertilizer Research*, 33:115-121.

Woomer, P.L. (1993), "The Impact of Cultivation on Carbon Fluxes in Woody Savannas of Southern Africa," in J. Wisniewski and R.N. Sampson (eds.) *Terrestrial Biospheric Carbon Fluxes: Quantification of Sinks and Sources of CO_2*, Kluwer Academic Publishers, Netherlands.

CHAPTER 13
RANGELANDS AND GRASSLANDS

13.1 INTRODUCTION

Rangelands are lands generally characterized by low and/or erratic precipitation, poor drainage, rough topography, and often low soil fertility.[1] They occupy some 47% of the earth's surface (Heitschmidt and Stuth 1991). Fire, rainfall, soil type, and grazing animals are driving forces determining plant species composition, distribution, and productivity. Management of rangelands ranges from nomadic pastoralism, to subsistence farming, to commercial ranching. Many rangelands have been used the same way for thousands of years, while others have a relatively short history of use.

Carbon cycling and productivity in rangeland ecosystems are directly related to the amounts and seasonal distribution of precipitation and only secondarily controlled by other climate variables and atmospheric chemistry. Most carbon storage in grasslands, savannas, and deserts is below ground. Estimates done using the soil/plant simulation CENTURY model (Parton *et al.* 1987, 1992) suggest that 560 Pg C is stored in biomass and litter, while 1100-1400 Pg C is stored in roots and soils of the terrestrial biosphere, with the carbon in grasslands and savannas estimated at 417 Pg C (Sampson *et al.* 1993). However, rangeland productivity may vary as much as five-fold because of timing and amounts of precipitation (Walker 1993). Non-sustainable land use practices such as inappropriate plowing, overgrazing of domestic animals, and excessive fuelwood use are the root causes of the degradation of rangeland ecosystems (Ojima *et al.* 1993; Sampson *et al.* 1993). It is estimated that about 70% of the world's drylands are at least moderately degraded (Dregne *et al.* 1991).

Methane production by wild and domestic animals has been estimated to be around 80 Tg/yr (Cicerone and Oremland 1988) or about 15% of the world's total methane emissions. N_2O is produced as part of the N cycle and is subject to increases with alteration of the N cycle through land use changes (Ojima *et al.* 1993, Mosier *et al.* 1991). Temperate grasslands of the world are estimated to contribute 0.1 Tg N_2O annually or about 0.65% of the world's total sources of N_2O (Matson and Vitousek 1990).

13.2 OVERVIEW OF MITIGATION OPTIONS

Reduction of GHG emissions in the rangelands sector primarily involves the reduction of methane production by wild and domestic ruminant grazers, and increasing storage of carbon, which is dependent on improving rangeland health where needed. The dynamics of soil organic matter is one of the major indicators of change in greenhouse gas fluxes in grassland and rangeland ecosystems. Altering land use practices can influence the rate of changes in soil organic matter (SOM) levels and be used to mitigate efflux of CO_2 into the atmosphere. Good soil management is the key to maintaining or increasing carbon storage and protecting rangeland health. Improving rangeland health, and thus the amount and kind of vegetation, will also reduce methane emission from ruminant animals by improving the quality of their diet. Research suggests that small reductions in the number of grazing animals (stocking rate) or altering the timing of grazing could result in large soil sinks for atmospheric CO_2 (Metherell *et al.* 1993) and in methane emission reduction (Howden 1991, Galbally *et al.* 1992).

Table 13-1 (found at the end of this chapter) presents a list of options to improve rangeland health and also mitigate GHG emissions. Not all the practices listed are relevant to every country, social system, or rangeland type. The suggested practices will only be successful if local communities benefit from their implementation. None of the practices is likely to significantly improve rangeland health and carbon

---

[1] The term rangelands as used in this chapter includes grasslands and drylands.

sequestration without adequate amounts and appropriate timing of rainfall. The most successful rangeland improvement projects for unhealthy rangelands were implemented in years of high rainfall (Heady 1988).

13.3 ANALYTICAL APPROACHES FOR RANGELANDS MITIGATION ASSESSMENT

There are two main approaches which can be used to analyze options for reducing GHG emissions or increasing carbon storage on rangelands. The first involves case studies based on existing rangeland management practices.

The CENTURY ecosystem model (Parton *et al.* 1987 and 1993) can be used to evaluate the effectiveness of various mitigation options for increasing or maintaining stocks of SOM and improving rangeland health in grassland and rangeland ecosystems. In developing simulations, data will be needed for weather, soil characteristics and spatial distribution of grassland and rangeland. Land use management data will be needed in order to simulate reasonable scenarios of current and alternate land use practices or mitigation options. In developing the mitigation options, identification of the key input and output variables (such as soil C levels, livestock production, plant productivity, N gaseous losses, etc.) is critical. The determination of specific mitigation practices that are appropriate for a particular situation needs to be identified and described in enough detail to be implemented in the model framework.

A key step in the initial phase of analysis is to describe the physical, biological, and the human-related characteristics of the study area that identify the critical factors affecting C storage and fluxes, and fluxes of other GHGs. The physical variables include information related to the climate and soil systems. The climate data to be defined for a site or region include the monthly average maximum and minimum air temperature and the monthly precipitation. Climate data collected over a range of years to capture the interannual and the decadal variations in weather patterns is desirable. The soil data needs to include information on soil texture, water holding capacity, soil pH, bulk density, and depth. Other physical data that are useful to have include topography and parent material type.

Biology information includes description of the vegetation, disturbance patterns, seasonal dynamics, and nutrient cycling. The vegetation information needs to include a general description of functional attributes of the components of the ecosystem, such as what percent biomass is contained in each category. What seasonal pattern of growth and mortality takes place in the different functional categories? What physiological and biogeochemical attributes does the dominant vegetation possess? For instance, data on the lignin content of plant material, plant C and N content, and the dominant photosynthetic pathway would be useful. In addition, if information on the current, and if available, on historical levels of soil C and N levels were available, analysis of changes in these components would be better documented.

Human activities affecting soil C and other GHG fluxes include land use management related to grazing, fire, animal residue, cropping patterns, harvest techniques, tillage practices, and fertilizer use. Modification in these land management practices provides the techniques to implement different mitigation options to reduce GHG emissions and increase C storage in grassland and rangeland ecosystems. Development of appropriate mitigation options will be determined by specific economic, political, and cultural characteristics of the situation that is being studied.

A second analytical approach involves a comprehensive assessment of the rangeland sector and its role in the country's formal and informal economy, including that of providing climate change mitigation. Analysis techniques which link ecosystem process studies across a range of environmental factors, such as climate, soils, land use, in a geographically explicit manner are currently being used. This analysis structure allows one to assess the regional impact of global environmental changes and the effectiveness of different mitigation options on soil carbon storage and other ecosystem feedbacks to the atmospheric system (e.g. H_2O, CH_4, and N_2O). This structure takes information from ecosystem process studies and from spatial data on soils, climate, land use, and policy-relevant information to make assessments of the impact of changing management practices for greenhouse gas mitigation.

A comprehensive assessment consists of the following elements:

- inventory of current rangeland area, vegetation, soil types, and demand for forage, fuelwood, or other uses;
- evaluation of the current condition (health) of ecosystem types;
- assessment of rangeland ecosystem change;
- assessment of future land area available for domestic grazing animals and wildlife, given the demand for land by other sectors;
- assessment of future demand for forage, fuelwood, agriculture, or other uses from rangeland ecosystems;
- projection of the land areas as well as the livestock and wildlife production under both baseline and mitigation scenarios;
- estimation of the potential for reducing GHG emissions and/or sequestering carbon for each option considered;
- estimation of the costs and non-GHG benefits of each option;
- based on the above analysis, identification of potentially attractive mitigation options;
- estimation of the potential carbon sequestration or GHG reduction for each mitigation option; and
- documentation of policies, institutional arrangements, and incentives necessary for the implementation of options.

In this approach, the analyst should examine the rangeland sector within the context of land-use demand by all sectors (see Chapter 10). Such a comprehensive approach will result in identifying a mix of practices which use the fewest sources (bio-physical, cultural, and economic) and interfere the least with established viable pastoral practices while aiding the mitigation of climate change. This approach will allow for a cost-effective implementation of a subset of the options depending on the individual country, pastoral system, local culture, and available land resource. This approach, which is similar to that for the forest sector, also reduces the possibility of double-counting of GHG flows, costs, and benefits.

13.4 DEVELOPING A BASELINE SCENARIO

The first steps are to use available maps and/or data to determine the location and amounts of rangeland vegetation, and to identify the existing patterns of livestock and wildlife use.

13.4.1 Assessing Rangeland Health

A baseline assessment of the ecosystem should lead to the categorization of rangelands as either healthy, at risk, or unhealthy, each providing a different potential for mitigation options. The evaluation of any option or practice will be undertaken against the background of this baseline assessment.

Rangeland health is defined as the degree to which the integrity of the soil and the ecological processes of rangeland ecosystems are sustained (NRC 1994). Evaluation of rangeland health will provide the initial assessment of opportunities for increasing carbon sequestration on rangelands. No one factor is sufficient to evaluate rangeland health; three criteria are suggested (NRC 1994):

- **Degree of soil stability and watershed function**. Soil stability is directly affected by soil erosion through wind and water. Breakdown of soil structure, reduced infiltration and thus water holding capacity, reduced organic matter and nutrient cycling, compaction, reduced germination, change in species composition, and productivity are all potential ecosystem effects caused by increased soil erosion.

  The evaluation of soil erosion can be based on several environmental factors such as the presence of the A-horizon, rills and gullies, pedestaling, scour or sheet erosion, sedimentation fans, or dunes.

- **Evaluation of the integrity of nutrient cycles and energy flow**. The capacity of rangelands to provide goods and services to human populations depends on the ability of plants to capture sunlight through photosynthesis and on the accumulation and cycling of nutrients over time. Interruption of nutrient cycles through poor land-use practices can lead to degradation of rangeland vegetation, reduction in productivity, and thus changes in ecosystem potential for carbon sequestration.

  Indices that can be used to evaluate ecosystem integrity in nutrient cycling and energy flow include the distribution of plants (growth forms, life-forms, and species), the degree of fragmentation in litter distribution, rooting depth, and community distribution.

- **The presence of functioning recovery mechanisms**. Useful indicators of the ecosystem resilience might include increasing plant cover, increasing plant vigor, changes in kind and number of seedlings, changes in plant age-class distribution, and other community attributes which would lead to greater soil stability and improved nutrient storage and cycling.

13.4.2 Assessing Rangeland Ecosystem Change

Linear response to grazing pressure and fluctuation in rainfall has been assumed by early successional models of rangeland vegetation change. Under the succession model, soil productive potential remains constant and plant community composition and productivity changes in a predictable inverse linear fashion in response to grazing or drought, i.e., remove grazing animals and health improves, add animals and health declines.

Rangelands generally do not respond as the successional model predicts because annual variation in amounts and distribution of rainfall already results in a wide variation in annual productivity (Walker 1993). Coupled with fire, grazing, and human activities, and a combination of lag effects in community response (thresholds which are not easily re-crossed), and multiple trajectories, there appears to be a confusing array of potential community states.

A new conceptual model of rangeland ecosystem functioning was proposed by Westoby *et al.* (1989). The model requires a knowledge of the changes that can take place in a particular rangeland and what causes them. Its value lies in making the understanding of community dynamics explicit. The most significant advantage of using this model to understand rangeland ecosystem dynamics is the recognition that the future may contain an array of possible states, dictated by an oncoming stream of events, and presenting a mixture of opportunities and hazards. The objective of rangeland management then is to seize the opportunities and as much as possible avoid the hazards (Walker 1993).

With this framework in mind, countries can evaluate existing rangeland condition (health), catalogue existing and potential states, catalogue management and natural events that cause transitions from one state to another, and evaluate which states provide the greatest opportunity for increasing carbon sequestration and/or reducing GHG emissions.

Particular combinations of rainfall, topography, soil type, and biological composition give rangelands individual characteristics, which, when coupled with different kinds of human use, make a state-transition model a useful conceptual tool for evaluating potential ecosystem response and thus the opportunity for carbon sequestration.

13.5 ANALYSIS OF MITIGATION OPTIONS

13.5.1 Potential Mitigation Practices

Table 13-1 identifies the kinds of practices which affect livestock and wildlife populations and the rangelands upon which they graze. If implemented, these practices will have different effects on reducing methane or increasing carbon storage. Different rangelands and social systems will require application of different mixes of practices. Although ecological, social, and economic costs and benefits are qualitatively addressed, issues of biodiversity and genetic conservation of plants and animals for future uses are only hinted at. The evaluation of different mitigation practices gives rise to different issues as shown in the following classes of mitigation options:

- Rehabilitation of degraded rangelands offers a very attractive opportunity to sequester carbon. The most likely practices under this option include afforestation, reforestation, grass and shrub establishment, control of grazing lands, halophyte establishment on salinized lands, etc. (Glenn *et al.* 1992). In an assessment of costs of rehabilitating the degraded rangelands of the world by applying a combination of the above practices, UNEP (1991) estimated that it will cost about US$5-8.8 billion per year over a period of 20 years. The CENTURY model was used to estimate the carbon flux difference between the baseline and the sustainable management scenario; the estimated benefit was about 0.7 billion tons of carbon per year (Ojima *et al.* 1993). This amount is equivalent to a cost of US$10 per ton of carbon sequestered, which is comparable or superior to the estimates in the forestry sector.

  The desirability of this mitigation option is further enhanced by the fact that most of the carbon sequestration in rangelands is through soil storage, which has a half-life of hundreds of years—a period much longer than that for carbon sequestered in above-ground biomass. The process of undertaking cost-benefit analysis for rangeland rehabilitation is similar to that described in the forestry chapter.

- The viability of reducing livestock numbers will depend on the value of livestock to local and national economies and the value of animals as a social resource. For example, in sheep grazing systems in Australia, Howden (1991) found linear relationships between methane and nitrous oxide emissions and stocking rate. Using DYNAMOF, a simulation model, they showed that reducing emissions by 20% without any technological changes required an 18% reduction in stock numbers with a 15%-17% reduction in net cash income. However, using GRASSMAN, an agricultural decision-support model, Howden showed that changing the time of lambing, reducing stock in overgrazed areas, and managing fire frequency led to a significant reduction in GHG emissions without substantial effect on net income.

- Changing the mix of animals depends on the kind of rangeland and proposed mix of animals. If a country is considering only cattle and small stock (sheep or goats, etc), the mix may not be ecologically efficient; rather it may reflect an economic risk aversion where in bad times cattle die but goats survive. In this kind of grazing mix, the ecosystem may deteriorate. A mix of cattle and wildlife ruminants may be both ecologically and economically efficient.

- Changing animal distribution through salt placement, development of water sources, or fencing can increase carbon sequestration through some increase in plant cover and improved health of the root system through lighter intensity of grazing. However, none of the changes in animal distribution is expected to affect methane production.

- The practices most likely to reduce methane emissions from domestic and wild ruminants involve improving the quality of the diet. Providing protein supplements is one alternative. However, increasing native grasses and planting other adapted, productive species (such as halophytes where appropriate) will provide additional benefits to local communities and economies.

- Other practices like application of herbicides, use of mechanical methods to rehabilitate unhealthy rangeland, and watershed scale developments involve greater ecological, social, and economic costs. Specific values are not given in Table 13-1 because they vary depending on specific goals, rangeland health, and country. Economic costs can be derived from Heady (1988), Valentine (1990), and Child *et al.* (1987).

13.5.2 Physical Criteria for Evaluating Mitigation Options

The nature of each mitigation practice will determine the type of physical criteria needed to evaluate the option. For the rehabilitation practices, the area in which the option can be implemented every year for the duration of the practice should be identified. The sum of the annual areas will form the maximum area available for the mitigation option. The schedule of implementation will be dictated by the size of the area and practical considerations regarding range and livestock management. Quite possibly, some of the above mentioned practices can be carried out on the same land area with cumulative effect on emission reduction and/or carbon sequestration. In general, the following physical criteria should be estimated:

- **Land Availability**. The land on which each mitigation option would be implemented should be identified. The schedule of possible implementation should be used to estimate the total area available for the practice.

- **Emissions Reduction and/or Carbon Sequestration**. Carbon flows associated with each option at all stages of the project, including emissions from biomass burning, salvage, use of wood products, and emissions from soil disturbance should be estimated. Other trace GHGs such as CH_4, N_2O, NO_x, and NMHC should be estimated if the option in question will avoid the emission of significant quantities. In most of the options listed above, a substantial decrease in methane emissions is expected, and this should be estimated using known figures for methane emission per body weight of an animal.

Estimates of uptake by vegetation and soils within the area should be done on the basis of net primary productivity of the woody biomass, including the net storage in soil and detrital material.

Applying these estimates to the total area available for each specific option, one can obtain the total carbon sequestration or emission reduction for the option.

13.5.3 CENTURY Model Application for Rangeland Mitigation Assessment

CENTURY is a general ecosystem model which simulates plant-soil dynamics of grasslands, forests, crops and savannas. The model runs using a monthly time step. The major input variables for the model include: (1) monthly average maximum and minimum air temperature; (2) monthly precipitation; (3) lignin content of plant material; (4) plant C and N content; (5) soil pH and soil texture; (6) atmospheric and soil N inputs; and (7) initial soil C and N levels.

The model has three major submodels. These include:

- a biophysical submodel which calculates hydrological and temperature drivers;
- a plant production submodel which calculates above- and below-ground plant processes; and
- a soil organic matter submodel that calculates changes in soil C and N and plant detritus.

The biophysical submodel is used to determine rate constants of decomposition, potential and actual evaporation and transpiration rates, and plant growth and death rates derived from climate and soil properties of a site or region. Inputs of monthly average maximum and minimum air temperature, monthly precipitation, and soil texture are used to determine soil temperature, soil water balance and other biophysical control factors for ecosystem dynamics.

The plant production and the soil organic matter (SOM) submodels are used to evaluate changes in C levels for different plant and soil components of the ecosystem. Units of these plant and soil pools of C are expressed as grams C per unit area (i.e., $g\ C / m^2$). Additional functions are used to simulate various land use management or disturbance regimes, such as forest planting, cropland conversion, fire, or grazing intensity and frequency. Appendix 13-1 provides further description of the CENTURY model and its use in simulating fire and grazing effects in grassland and rangeland ecosystems.

13.5.4 Evaluation of Economic Criteria

The options in the rangelands may require a gradual transition from the current rangeland health to a new sustainable equilibrium. As the state of the range improves, the carrying capacity will also change, and there will be a change in the flow of GHG associated with each state. The economic evaluation requires one to track the value of costs and non-GHG benefits over the transition period and compute the discounted value. Once the health of the range is restored, and the optimal mix of animals and diet quality is determined, the net change in GHG and value of resources needed can be assumed to continue in perpetuity. The net present value (NPV) of the perpetual stream of net benefits can then be compared to the corresponding net emission reduction.

- **Physical Inputs and Outputs.** The first step is to identify and quantify all necessary physical inputs required for implementation of each option during the initial operations, management, harvesting (if applicable), etc. These should include estimates of land, labor, equipment and material needed to support the project or option throughout its lifetime. For all options, one also has to identify constraining factors such as expertise, technology, and capital investment because these factors may affect both the cost as well as the implementation possibility of the option.

  Together with the physical inputs, one has to estimate the physical output in terms of desirable products like number of animals, wool, milk, and other products as applicable.

- **Unit Costs and Benefits.** For each of the physical inputs, one has to estimate the unit cost at the time of use. For each of the physical outputs of desirable products, an estimate of its price is necessary.

  The costs of a mitigation option include (1) the present value of the stream of expenses sufficient to cover the project's planning, development, occasional and recurrent expenses, and (2) the present value of the project's opportunity cost. Other cost components such as land rental (opportunity costs), maintenance, and monitoring and evaluation may need to be included if applicable. The opportunity cost is important since it captures the benefits derived from land use in the absence of a mitigation option, given the current land-use patterns. Opportunity cost may be evaluated using various methods, depending on the land in question and the likelihood of producing various goods and/or services if it is not used for the given option. These approaches include land rent, land market price, and net benefits obtainable from an alternative land use. In all these cases, land values and benefits from alternative use should be adjusted to account for existing significant price distortions due to subsidies, zoning regulations, etc.

  In addition to GHG impacts, the implementation of a mitigation option will result in other monetary and non-monetary benefits. Direct benefits may include goods such as livestock and services such as recreation. Indirect benefits may include such items as employment for local inhabitants and watershed protection.

13.5.4.1 Cost-effectiveness criteria

- **Initial Cost per ha and per tC.** This indicator provides useful information on the amount of resources required at the beginning to establish the project.

- **Endowment Requirements per ha and per tC.** The requirements are the sum of initial cost and the discounted value of all future investment and recurring costs during the lifetime of the project. For projects which do not have substantial monetary benefits, this indicator is quite useful because it provides the endowment necessary to maintain the project in perpetuity.

- **Net Present Value (NPV) per ha and per tC.** This indicator provides the net direct benefit to be obtained from the project. For most range improvement options as well as animal-based practices (e.g., animal mix change), the NPV indicator is expected to be positive for the option to be attractive. A benefit:cost ratio of all discounted values also provides a good measure of profitability.

- **Benefit of Reducing Atmospheric Carbon (BRAC).** This indicator expresses the NPV of a project per unit of atmospheric carbon reduced by a mitigation option. In the case of other GHGs, this indicator can be computed on the basis of CO_2-equivalent GWP (see Chapter 2). The formulation of the indicator varies with the rate at which economic damage might increase, and it allows time-dependent evaluation of atmospheric carbon as may be deemed necessary.

- **Imputed and Non-monetary Costs and Benefits.** After compiling the criteria given above, all the identifiable costs and benefits which one is currently unable to evaluate should be listed for each mitigation option. Imputed values should be listed separately from the direct costs and benefits. The intangible benefits and costs should also be listed for each mitigation option. To the extent possible, one should identify the likely bearers of costs and benefits, including the non-monetary items. Although there may not be a coherent method for comparing the non-monetary and intangible costs and benefits associated with each option, their enumeration helps the policy-makers in the choice and implementation of the various mitigation options.

13.6 CONSTRUCTING A MITIGATION SCENARIO(S)

Having compiled physical, economic, and other information for each mitigation option, one can construct mitigation scenarios that show the aggregate impact of selected options. The impact of each mitigation option is measured against the baseline scenario developed for rangeland ecosystems.

One useful way to summarize the results is a supply curve for emissions reduction or carbon sequestration. For example, the initial cost per ton of carbon stored can be used to plot a curve which shows the amount of carbon that could be stored at increasingly higher initial cost. The other indicators like endowment requirement or NPV or BRAC could also be used to plot similar curves.

To construct a cost/supply curve, the unit values (cost and carbon storage per hectare) are combined with the area available for each option to obtain estimates of total emissions and costs of each option. Methane abatement can be treated in a similar fashion, either in terms of tons of methane or CO_2-equivalent GWP.

13.7 POLICY CONSIDERATIONS

It is important to recognize that rangelands are variable and episodic, and that vegetation response is not linearly related to change in livestock number. Thus, risk management strategies would include long-term, low stocking, or changing animal numbers annually and tracking annual variation in precipitation. Either approach requires flexible management response to different events, opportunities, and hazards.

Another policy consideration is that although rangelands have historically been used for livestock production for meat, wool, hides, milk, blood, and/or pharmaceuticals, an equally important objective may be to maintain the maximum number of animals as a social resource. Rangelands are also increasingly affected by human activities for mineral production, construction materials, fuel, and chemicals. Additionally, rangelands provide habitat for wildlife, threatened and endangered species, anthropological sites, and recreational activities. As the human population grows, rangelands also incur increased demand for marginal agriculture production. All these activities and uses potentially affect rangeland health and thus the potential of the ecosystem to sequester carbon.

Table 13-1. Practices to Improve Rangeland Health and Mitigate GHG Emissions.*

Practice	Healthy rangeland	Unhealthy rangeland	Carbon	Methane	Bio/physical benefit/cost	Social/cultural benefit/cost	Economic benefit/cost	General comments
Reduce animal numbers (in animal unit months (AUMs)	No	Yes	Increases carbon sink because of increasing vegetation cover and better root growth	Reduces animal methane production through reduction in total number	Increases plant cover, increases soil organic matter, and improves productivity	Depends on country and value of animals as a social resource	Depends on value of livestock products to national and/or local economy	Positive eco- system effect if sufficient rainfall. May require alternative sources of local food support, thus changes in food production policies
Change mix of animals	Yes	Yes	Possible increase in carbon sink with change in plant species	No known effect	Potential changes in plant species composition	Depends on country and cultural value of specific animal type	Depends on value of livestock products	Positive effect in general, improves efficiency of utilization
Alter animal distribution by placement of salt	Yes	Yes	Increases carbon sink because of increasing vegetation cover overall	No effect	Useless in rangeland areas already high in salt	Appropriate in countries where animals graze extensively rather than being herded	Cost of salt and distribution of salt	Positive. Not applicable for herding systems
Alter animal distribution by placement of water sources	Yes	Yes	Increases carbon sink because of increasing vegetation cover overall	No effect	Developed water resources may not be sustainable. Potential cost to long-term productivity	May affect territorial and property boundaries	Motorized water sources are often too costly to purchase or maintain	Negative if used to increase numbers of animals. Positive if used to alter animal distribution

* The possible effect of implementing a practice is given for carbon and methane, and qualitative cost/benefits estimates are provided. Unhealthy rangelands are those lands where soil loss, plant species and cover loss, species invasions, and interrupted and poorly functioning nutrient cycling are the norm. Healthy rangelands, on the other hand, have nutrient cycling and energy flows intact, soils are not eroding, and plant species composition and productivity is indicative of a functioning ecosystem.

Table 13-1. Practices to Improve Rangeland Health and Mitigate GHG Emissions - continued *

Practice	Healthy rangeland	Unhealthy rangeland	Carbon	Methane	Bio/physical benefit/cost	Social/cultural benefit/cost	Economic benefit/cost	General comments
Alter animal distribution by placement of fences	No	Yes	Increases carbon sink because of increasing vegetation cover overall	No effect	Benefit is to control domestic animal number and distribution	Depends on country and livestock/ wildlife system. Costs potentially outweigh benefits	Varies depending on country and source and kind of materials	Potentially interferes with wildlife migration
Provide livestock protein supplement	Yes	Yes	No effect	Decrease methane production	Perhaps will reduce extensive grazing to some degree	Possible where animals are herded	Cost of protein blocks or similar supplement	Potentially difficult to distribute to local areas
Increase native grasses and or plant adapted species	No	Yes	Increases carbon sink because of increasing vegetation cover overall	Possible benefit of methane reduction by increasing quality of diet	Benefit in retention of native species for gene conservation	Local people rely on native species for medicine and other health- related goods	Depends on the value of the livestock and wildlife products, and value of herbal medicine	Potential unknown benefits from native species. Adapted species survive in long term
Selective application of herbicides	No	Possibly	Potentially increase carbon sink	Potentially increase if expand animal numbers	Cost if non-target species, pollution of water, damage to food chain	Cost if non-target species, pollution of water, damage to food chain	Varies depending on country and source of herbicide	Cost if non-target species, pollution of water, damage to food chain
Mechanical treatment or restoration	No	Possibly	Potentially increase carbon sink	Potentially increase if expands animal numbers	Potential for large-scale alteration of soil and vegetation	May not fit pastoral system	Varies with country depending on availability of equipment	Benefit depends on success of treatment relative to disruption of ecosystem

Table 13-1. Practices to Improve Rangeland Health and Mitigate GHG Emissions - continued *

Practice	Healthy rangeland	Unhealthy rangeland	Carbon	Methane	Bio/physical benefit/cost	Social/cultural benefit/cost	Economic benefit/cost	General comments
Plant halophytes (salt tolerant-species)	If appropriate	If appropriate	Increase carbon sink and increase productivity	No known effect	Benefit with increased plant cover and productivity	Benefit with increasing forage production for livestock and wildlife	Cost of planting and maintaining with irrigation	Brings into production otherwise non-productive land
Apply prescribed burning	Yes	Yes	Increase carbon sink and increase productivity in the long term on appropriate rangeland types	Possible benefit of methane reduction by increasing quality of diet	In systems adapted to fire, can increase productivity, maintain nutrient cycling	Use of fire can be part of social system. Utilizes local knowledge	Threat of wildfire and destruction of resources	Short-term increase in CO_2 to atmosphere, long-term benefits in adapted systems
Implement agroforestry systems	Yes	Yes	Increase carbon sink and increase productivity in the long term on appropriate rangeland types	Possible benefit of methane reduction by increasing quality of diet	Possible benefit with increased plant cover, diversity, and productivity	Benefit with increasing forage production for livestock and wildlife	Cost of planting and maintaining	Increases carbon storage in trees. Benefit in diversity and productivity if adapted species
Develop large-scale watershed projects	Possibly	Possibly	Increase carbon sink and increase productivity	Benefit, methane reduced by increasing quality of diet	Potential for large land disturbance, with benefit to human and animal populations because of regulated and regular water supply	Potential for improved food production, both plant and animal	Cost of dams etc, benefit hydro-electric power	Potential for increased human and animal populations because of increase in water

REFERENCES

Child, R.D. *et al.* 1987. Arid and Semiarid Rangelands: Guidelines for Development. Winrock International Institute for Agricultural Development. 291 p.

Cicerone, R.J. and R.S. Oremland. 1988. Biogeochemical aspects of atmospheric methane. *Global Biogeochemical Cycles* 2(4):299-327.

Crutzen, P.J., I. Aselmann, and W. Seiler. 1986. Methane Production by Domestic Animals, Wild Ruminants, Other Herbivorous Fauna and Humans. Tellus 38B:271-284.

Dregne, H., M. Kassas, and B. Rosanov. 1991. Desertification Control Bulletin 20, 6-18, UNEP. Nairobi, Kenya.

FAO (Food and Agriculture Organization) (various years), *FAO Production Yearbook*, FAO, United Nations, Rome, Italy.

Galbally, I.E., P.J. Fraser, C.P. Meyer and D.W.T. Griffith. 1992. Biosphere/atmosphere exchange of trace gases over Australia. Australia's Renewable Resources: Sustainability and Global Change. Bureau of Rural Resources, Australia. p. 67-116.

Glenn, E.P., L.F. Pitelka, and M.W. Olsen. 1992. The use of halophytes to sequester carbon. *Water, Air, and Soil Pollution* 64:251-263.

Heady, H.F. ed. 1988. The Vale Rangeland Rehabilitation Program: An Evaluation. Resource Bulletin. PNW-RB-157. USDI Bureau of Land Management, USDA Forest Service, Portland, OR. 151 p.

Heitschmidt, R. and J. Stuth (eds). 1991. Grazing Management: An Ecological Perspective. Timber Press. Portland, OR. 259 p.

Howden, Mark. 1991. Methane production from livestock. Draft Australian Greenhouse Gas Emissions Inventory 1987-1988. Greenhouse Studies 10 (DASET):15-22.

Matson, P.A. and P.M. Vitousek. 1990. Ecosystem approach to a global nitrous oxide budget. *BioScience* 40(9):667-672.

Metherell, A.J., C.V. Cole and W.J. Parton. 1993. Dynamics and interactions of carbon, nitrogen, phosphorous and sulphur cycling in grazed pasture. Proc. XVIIth International Rangeland Congress. Palmerston North, New Zealand.

Mosier, A., D. Schimel, D. Valentine, K. Bronson and W. Parton.1991. Methane and nitrous oxide fluxes in native, fertilized and cultivated grasslands. *Nature* 350:330-332.

National Research Council (NRC). 1994. Rangeland Health. National Academy Press. Washington D.C. 180 p.

Ojima, D. , Parton, W.J., Schimel, D.S., and J.M.O. Scurlock. 1993. Modeling the effects of climatic and CO_2 changes on grassland storage of soil C. *Water, Air, and Soil Pollution* 70:643-657.

Parton, W.J., D.S. Schimel, C.V. Cole and D.S. Ojima. 1987. Soil Science Society of America Journal. 51: 1173-1179.

Parton, W.J., B. McKeown, V. Kirchner, D. Ojima. 1992. CENTURY *Users' Manual*, Natural Resource Ecology Laboratory, Colorado State University, Fort Collins, CO.

Sampson, R.N., M. Apps, S. Brown, *et al.* 1993. Workshop summary statement: terrestrial biospheric carbon fluxes--quantification of sinks and sources of CO_2. *Water, Air, and Soil Pollution* 70:3-15.

UNEP, 1991. UNEP Governing Council Decision - Desertification Control Bulletin 20, 3-5. Nairobi, Kenya.

Walker, B.H. 1993. Rangeland ecology: understanding and managing change. *Ambio* 22(2-3): 80-87.

Westoby, M., B.H. Walker, and I. Noy-Meir. 1989. Opportunistic management for rangelands not at equilibrium. *J. Range Management.* 42:266-274.

Valentine, J.E. 1990. Grazing Management. Academic Press, Inc. Berkeley, CA. 533p.

APPENDIX 13-1
THE CENTURY ECOSYSTEM MODEL

The plant production submodel simulates grass growth and includes the impact of grazing and fire on plant production. Potential plant production is calculated as a function of soil temperature, available water and a self-shading factor. The effect of moisture on plant production is a function of the ratio of current monthly precipitation plus the previous month's stored soil water to the potential evapotranspiration rate. The soil water holding capacity also influences plant production by modifying the amount of stored soil water so that lower water holding soils (e.g., sandy soils) have a higher growth rate under dry conditions. The potential plant production rate is reduced if there is insufficient N. A maximum and minimum C to N (C:N) ratio is specified for new production of live roots and shoots. The model calculates root/shoot ratios as a function of the annual rainfall.

The SOM submodel simulates the dynamics of C and N in the organic and inorganic parts of the soil system. The flow diagram for soil C shows that soil C is divided up into three major components which include active, slow, and passive soil C. Active SOM includes live soil microbes plus microbial products (the total active pool is approximately 2 to 3 times the live soil microbial biomass); the slow pool includes resistant plant material (for instance, lignin-like components) and soil-stabilized plant and microbial material, while the passive material is very resistant to decomposition and includes physically and chemically stabilized SOM. The flows of C are controlled by the inherent maximum decomposition rate of the different pools and the water and temperature-controlled decomposition factor. Average monthly soil temperature at the soil surface controls the temperature function, and the ratio of stored water (0-30 cm depth) plus current monthly precipitation to potential evapotranspiration is the input for the moisture function. Microbial respiration occurs for each of the decomposition flows. The partitioning of decomposition between stabilized SOM and CO_2 flux is a function of soil texture for the stabilization of active C into slow C (increasing CO_2 flux for sandy soils and less soil C storage).

Plant residues (shoots and roots) are partitioned into structural (resistant to decomposition) and metabolic (readily decomposable) plant material as a function of the initial residue lignin (L)-to-nitrogen (N) ratio. The lignin fraction is assumed to be part of the structural material and (i.e., L:N ratio) controls the decomposition rate of structural material.

The organic-N flows follow the C flows and are equal to the product of the carbon flow and the N:C ratio of the state variable that receives the C. The C:N ratios of the soil state variables receiving the flow of C are a function of the mineral N pool (NO_3^- plus NH_4^+) and vary within the ranges 3-15, 12-20, and 7-10, respectively, for active, slow, and passive SOM. The C:N ratio of newly formed surface microbial biomass is a linear function of the N content of the material being decomposed and increases from 10 to 20 as the N content decreases from 2.0% to 0.01%. The C:N ratio of slow SOM material formed from surface microbes is equal to the C:N ratio of the microbes plus 5.0. N associated with carbon lost in respiration is assumed to be mineralized. Given the C:N ratio of the state variables and the microbial respiration loss, decomposition of metabolic residue, active, slow, and passive pools generally results in net mineralization of N, while decomposition of structural material immobilizes N. The model also uses simple equations to represent N inputs due to atmospheric deposition and N fixation and calculates N losses due to N_2, NO, N_2O, and NH_3 gas fluxes and NO_3 leaching.

Simulated fire and grazing effects in grassland and rangeland ecosystems. The major impact of fire is to increase the root to shoot ratio, increase the C:N ratio of live shoots and roots, remove vegetation and return nutrients during the years when fire occurs. The effect of different intensities of fire in herbaceous vegetation can be parameterized by specifying the fractions of live shoots, standing dead and surface litter removed by a fire along with the return of N, P, and S in inorganic forms.

Grazing removes vegetation, returns nutrients to the soil, alters the root to shoot ratio, and increases the N content of live shoots and roots. The model has three options for dealing with the impact of grazing on the system. First, a zero-direct effect option where there are no direct impacts of grazing on plant production except for the removal of vegetation and return of nutrients by the animals. Secondly, a linear negative impact option where grazing on plant productivity results in a linear decrease in potential plant production with increasing grazing intensity while holding root:shoot ratio constant. The third option employs an optimal grazing effect where plant production is increased for low to moderate grazing (grazing removal less than 35%) and decreases sharply for heavy grazing levels (>40% removed per month). The root:shoot ratio is constant for low to moderate grazing levels and decreases rapidly for heavy grazing levels. In all three options, the nutrient content of new shoot will increase in relation to the residual biomass.

The grazing options can be parameterized to remove defined fractions of above-ground live and standing dead plant material each month. The fractional returns of C, N, P, and S are specified, having allowed for losses in animal carcasses and milk, transfer of dung and urine off the area being simulated, volatile losses of N from dung and urine patches, and leaching of N and S under urine patches. The proportion of N, P, and S returned in organic forms is also specified as is the lignin content of the feces.

CHAPTER 14
WASTE MANAGEMENT

14.1 INTRODUCTION

This chapter presents methods for evaluating emissions-reduction options for the waste management sector. The primary sources of emissions in this sector include landfills and wastewater treatment. Options for mitigating emissions from livestock manure are discussed in Chapter 12.

Landfills. Methane is emitted from landfills as the result of the anaerobic decomposition of organic wastes. The methane migrates through the waste laterally and vertically, eventually escaping to the atmosphere. Landfills are a major global source of methane, contributing between 20 and 60 Tg of methane annually (USEPA, 1994). The major uncertainties in these estimates include the amount of organic material actually disposed of in landfills by different countries, the portion of the organic waste that decomposes anaerobically, and the amount of methane generated per unit of decomposed waste. About two-thirds of methane emissions from landfills come from the more developed countries of the world, another 15% from countries with transitional economies, and 20% from developing countries (USEPA, 1994).

The primary method for reducing methane emissions from existing landfills is to collect and combust the landfill gas. Diverting organic refuse to other disposal and treatment options and away from landfills can reduce future emissions.

Wastewater Treatment. Wastewater and sludge, its residual solids by-product, produce methane emissions if they are stored or treated under anaerobic conditions (in the absence of oxygen). In some cases this methane is collected and used or flared, but in most circumstances the methane produced is released to the atmosphere. Although data are very limited, current global estimates of methane emissions from the management of residential, commercial, and industrial liquid and water-carried wastes are about 20 to 25 Tg per year (based on calculations of the organic content of wastewater in different regions). These estimates are no better than $\pm$50%.

The amount of methane emitted depends on the organic loading in the wastewater (measured as biochemical oxygen demand, or BOD) and the extent to which the organic material degrades under anaerobic conditions. The majority of the methane emissions from wastewater are believed to originate in developing countries where domestic sewage and industrial waste streams are often unmanaged or maintained under anaerobic conditions without control of the methane (USEPA, 1993b). However, much uncertainty remains regarding the emissions rates from wastewater treatment conditions found in many developing countries. Consequently, the emissions reduction that could potentially be achieved is not well quantified.

The most effective technique for reducing these emissions is to implement effective aerobic wastewater treatment systems. Such systems produce other significant human health and environmental benefits and are consequently also desirable for reasons other than methane mitigation. It is also feasible to use anaerobic treatment systems and recover and combust the methane produced.

14.2 MITIGATION OPTIONS

14.2.1 Landfills

There are two approaches to reducing methane emissions from landfills: (1) the methane generated in landfills can be recovered and used to produce energy; and (2) the quantity of landfilled waste can be reduced through source reduction, recycling, and other waste-management practices. These approaches, summarized in Table 14-1 are presented briefly below (see USEPA [1993a] for a more detailed discussion of the options).

14.2.1.1 Recovery and use of landfill gas

Landfill gas-recovery technologies have been demonstrated and are in use in several countries, and there is great potential for expanding these technologies in both developed and developing countries. The landfill gas is extracted through a series of wells drilled into the refuse. Negative pressure is applied to the wells to extract the gas, which is collected through a system of pipes. The gas is then processed to remove water and various contaminants. Between 50 and 85% of the landfill gas generated can typically be recovered from landfills, with well-designed projects achieving almost complete gas recovery.

The landfill gas can be used as an energy source, which offsets the cost of recovering the gas. Options for using the gas include:

- **Electricity Generation and Co-generation.** The recovered methane can be used to power an electric generator, with the generated electricity used on-site or sold to others for use. The waste heat produced during electrical generation can also be recovered and used for local heating needs. Electricity generation requires relatively large amounts of landfill gas and is therefore suitable for larger landfills. Economic viability depends primarily upon the price at which the electricity can be sold. Historically, more than 50% of landfill projects worldwide have been used for electricity generation (Richards, 1989), including projects in the U.S., U.K., Germany, Brazil, India, and the Netherlands. The Global Environment Facility (GEF) Fund is planning to undertake a project in Lahore, Pakistan, that will recover over 14 million m^3 per year of landfill gas (over a period of five years) from a large sanitary landfill designed to handle 730,000 tonnes of waste disposed of annually by the city's 5 million inhabitants (USEPA, 1993b).

- **Medium Btu Gas.** Landfill gas can be used directly as a medium Btu fuel to provide heating, cooling, or steam for industrial processes. Recovered landfill gas is already profitably used as a boiler fuel and for other industrial and residential applications in a variety of countries, including the U.S., Brazil, South Africa, and Chile.

- **Natural Gas Supply.** Landfill gas can be processed to produce "pipeline quality" gas (over 95% methane) with minimal impurities by removing moisture, carbon dioxide (CO_2), and hydrogen sulfide (H_2S). The gas must also be compressed at a certain minimum pressure to be injected into a pipeline or distribution system. There are several such projects in the U.S. that provide gas to local gas distribution systems.

14.2.1.2 Alternative waste-management strategies

Other technical options applicable in the near and longer term can reduce the landfilling of wastes through source reduction and recycling of organic materials. Paper products, for example, comprise a significant portion of solid waste in developed countries (e.g., 40% in the U.S.) and a growing portion of solid waste in some urban centers in developing countries (typically 5 to 20%) (USAID, 1988; Vogler, 1984). Paper products can be recycled into a variety of products, and the markets for the recycled products are, in most cases, identical to those for virgin paper products. Waste paper-recycling processes range in technical complexity and include technologies as simple as hand-operated baling presses.

Table 14-1.
Summary of the Technical Options for Reducing Methane Emissions from Landfills

Considerations	Methane Recovery and Utilization	Alternative Waste Management Practices
Recovery/reduction techniques	Recovery wells Collection systems	Source reduction Incineration Composting
Gas use/combustion options	Electricity generation Natural gas supply Flaring	None
Availability	Currently available	Currently available
Capital requirements	Medium	Low/medium
Technical complexity	Medium/high	Low/medium
Applicability	Existing and new landfills Nearby gas use Capital- and technology-dependent	Widely applicable
Methane reductions[a]	50-90%	Up to 100% of future emissions[b]

a) Reductions that may be achieved at individual landfills.
b) Does not reduce emissions from existing landfills.

Source: USEPA (1993a)

Composting is another promising waste management option that limits methane generation by reducing the amount of waste landfilled. Composting is applicable in the near and longer term, particularly in developing countries where the organic and moisture contents of municipal wastes are often high. The economics of composting projects can be favorable if a market exists for the compost. Markets often depend on the demand for fertilizer and are generally favorable in arid regions and other areas where organic soil supplements are needed.

Incineration of wastes is increasingly used in developed countries to reduce quantities of landfilled wastes, often combined with energy recovery from the combustion process. The costs of incineration are justified based on the increasing costs of handling municipal solid wastes. While there is potential for this technology to expand in developed countries, there is a much lower potential in developing countries because the wastes are frequently too moist for economically viable operations (USAID, 1988).

The preferred approach for reducing methane emissions from landfills for an individual country will depend on past waste-disposal practices. Countries that have already disposed of a significant amount of organic material in landfills should consider landfill gas recovery and utilization to reduce the emissions from these existing landfills. Landfill gas recovery is the only feasible method of reducing these emissions. Countries that are only now beginning to dispose of organic wastes in landfills should minimize the amount of organic material placed in landfills. By minimizing such disposal, future emissions will be reduced.

Because some organic material will likely continue to be disposed of in landfills, all new landfills should be designed with gas-collection and utilization systems.

Reducing the landfilling of organic materials and landfill gas collection have benefits in addition to reduced methane emissions including the following:

- decreased safety hazards from the migration of potentially explosive methane beyond the landfill boundaries;
- reduced odor problems from landfills; and
- reduced emissions of air pollutants such as volatile organic compounds (VOCs) and air toxics that adversely affect air quality and human health.

14.2.2 Wastewater Treatment

Although the factors affecting methane emissions from wastewater treatment remain very uncertain, it is generally believed that providing effective aerobic wastewater treatment will help reduce emissions. If anaerobic treatment is preferred to aerobic treatment, capturing and using the methane produced during treatment would also reduce emissions. These two approaches are summarized as follows (additional description is provided in USEPA, 1993a):

- **Aerobic Treatment.** Aerobic treatment includes aerobic primary and secondary treatment and land treatment. Aerobic primary wastewater treatment is achieved by sustaining sufficient oxygen levels during the primary phase of wastewater treatment (i.e., in oxidation ponds), using controlled organic loading techniques or providing oxygen to the wastes through mechanical aeration. Aerobic secondary treatment consists of stabilizing wastewater by prolonging its exposure to aerobic microorganisms which are either suspended (due to mechanical aeration) or attached to a fixed bed or a rotating cylinder. Finally, land treatment involves applying wastewater to the upper layer or the surface of soil, which acts as a natural filter and breaks down the organic constituents in the wastewater.
- **Recovery and Utilization of Methane from Anaerobic Digestion of Wastewater or Sludge.** If the wastes are treated (digested) under controlled anaerobic conditions, the resulting methane and other gases can be recovered and utilized as an energy source to heat the wastewater or sludge-digestion tank, produce power in other parts of the plant, or sell to nearby homes, industrial plants, or utilities. Flares are frequently used as part of these operations to dispose of excess methane.

Although the costs and benefits of these mitigation approaches remain to be assessed, initial research has indicated that improved wastewater treatment has a variety of benefits in addition to reducing methane emissions, including:

- reduction in the risk of water-borne diseases (Loehr, 1984);
- reduced eutrophication of receiving waters, which can be caused by high levels of phosphorus and/or nitrogen;
- elimination of odors from standing wastewater; and
- production of treated wastewater and sludge for various uses (e.g., recharging ground water, irrigation, soil enrichment, production of potting mixes and topsoil, turf production and maintenance, and reclamation of disturbed lands).

In the past, the benefits of reduced disease risk and protection of receiving water quality have been the controlling factors in the design and deployment of wastewater treatment systems. Improved wastewater treatment is likely to be desired for these reasons in most cases where untreated wastewater is producing methane emissions.

14.3 OVERVIEW OF THE MITIGATION ASSESSMENT PROCESS

The following four steps are recommended for performing a mitigation option analysis.

Step 1: Develop Scenario Inputs. The purpose of this step is to prepare inputs, such as the number of landfills and the volume of waste stream targeted for diversion away from landfills. These will serve as inputs for the preparation of both baseline and mitigation scenarios.

Step 2: Identify Target Subgroups to be the Focus of the Emissions-reduction Effort and Refine the Emissions Estimates for the Subgroup. The purpose of this step is to focus the analysis on those portions of the GHG source that are amenable to control. The target subgroups are selected based on the applicability of the mitigation options for the source. A more detailed baseline emissions estimate is then developed for the target subgroups. This detailed baseline will be used to estimate the emissions reduction that can be achieved.

Step 3: Evaluate the Mitigation Options for the Subgroup. The purpose of this step is to evaluate the impacts that the mitigation options have on the emissions and other characteristics of the target subgroup.

Step 4: Develop Baseline and Mitigation-emissions Scenarios. Using the information generated in Steps 1, 2 and 3, baseline and mitigation emissions scenarions may be developed for landfills and wastewater treatment.

Section 14-4 summarizes the inputs needed for developing emissions scenarios. Section 14-5 summarizes Steps 2 and 3, identifying the target subgroups and evaluating the mitigation options. Section 14-6 summarizes developing the baseline and mitigation scenarios.

14.4 SCENARIO INPUTS

The IPCC/OECD emissions inventory guidelines identify the inputs needed to estimate methane emissions for 1990. Scenarios of future emissions, with and without the implementation of mitigation options, may be developed by forecasting the key variables that drive the emissions.

Methane emissions from landfills are driven principally by the amount of degradable organic material disposed of in landfills. Future emissions depend both on the amount of waste already placed in landfills and the amount placed in landfills in the future. Therefore, future waste disposal rates must be estimated to forecast future emissions. Future disposal rates will depend on the amount of waste generated and the portion of the waste placed in landfills. Per capita waste-generation rates have been found to be dependent on consumer preferences and industrial activity. Because per capita income has been found to be one predictor of waste-generation rates, the data presented in the IPCC/OECD emissions inventory method can be used to estimate future per capita waste-generation rates as per capita incomes increase. The implications of changes in landfilling practices should be considered in the baseline scenario if such changes are expected.

Like emissions from landfills, methane emissions from wastewater treatment are driven principally by the amount of organic material that is treated anaerobically. The IPCC/OECD emissions inventory method recommends that the emissions be estimated separately for municipal and industrial wastewater. The amount of organic material in municipal wastewater nationally may be estimated as the population times a per capita waste-generation rate in units of BOD (i.e., kg BOD per capita/day). Regional rates of BOD generation per capita are given in the IPCC/OECD method, which are adequate for assessing the rough magnitude of emissions. More precise locally derived data would be preferred.

Organic material in industrial wastewater may be estimated based on production in several key industries. As with municipal wastewater, the IPCC/OECD method presents factors for estimating organic material in industrial wastewater.

To estimate methane emissions, the portion of the municipal and industrial wastewater that is treated anaerobically must be estimated. There is very little data on this important parameter, and additional assessments are required to improve the basis for making the estimates. At a minimum, the wastewater generated should be allocated into aerobic and anaerobic major treatment categories, such as:

- Aerobic: open pits/latrines, shallow ponds, ocean discharge, river discharge, and aerobic primary and secondary treatment systems.
- Anaerobic: deep ponds, anaerobic digesters, and septic tanks.

For each of the categories, an estimate is needed of the uncontrolled methane emissions that result per unit of BOD treated. Field measurement data are lacking for this parameter as well. Consequently, field measurement programs are needed to quantify current emissions and emissions reductions achievable by switching among treatment regimes.

14.5 ANALYSIS OF MITIGATION OPTIONS

The analysis of options for reducing methane emissions from landfills can proceed in two directions: gas recovery and waste diversion from landfills. The following approach is recommended for evaluating gas recovery:

- **Identify Landfills at Which Gas Recovery is Attractive.** The purpose of this step is to identify the population of landfills at which gas recovery and use is feasible and economically attractive. Experts in landfill gas recovery and use should be consulted to define those landfill characteristics that make landfills attractive for gas recovery. Generally, landfills should be relatively large (e.g., have at least 1 million metric tonnes of waste in place) and should be able to support the drilling of wells into the refuse (i.e., the refuse and the soil should be stable and not saturated with water). The actual landfill-gas recovery plant size will depend on the gas production rate, costs of gas collection and utilization, and the value of the energy derived from the gas in the specific circumstance.

- **Obtain, to the Extent Possible, Data on the Individual Landfills, e.g., from Landfill Operators.** For this set of landfills, detailed information on waste characteristics should be obtained for purposes of developing more detailed estimates of methane emissions. Test wells at selected landfills would be useful for verifying the emissions estimates for this set of landfills.

- **Evaluate Gas Recovery Projects.** For the targeted set of landfills, the costs and benefits of landfill gas recovery projects should be assessed. The costs of recovering the gas may be estimated using engineering cost estimates, such as those in USEPA (1993c) or similar country-specific cost factors. The amount of gas expected to be recovered may be estimated as a portion of the estimated emissions (e.g., 75% of emissions). The revenue from using the collected gas should be estimated based on the quantity of gas collected and the local value of the gas. A discounted cash flow analysis can then be done to identify the cost or benefit per unit of landfill gas emissions avoided.

- **Estimate Emissions Reductions.** Based on the evaluations of the gas recovery projects, the extent to which landfill methane emissions can be reduced from the targeted landfills can be estimated. The cost of reducing emissions by various amounts can be estimated as well using the results of the discounted cash flow analysis.

In addition to evaluating opportunities for recovering landfill gas, options for diverting waste away from landfills can also be considered. The major alternatives of recycling, source reduction, composting, and incineration may be examined. The impact on future methane emissions would be estimated based on the amount of degradable organic carbon (DOC) diverted away from landfills over time. The general approach to this assessment would include the following:

- Identify the component of the waste stream targeted for diversion away from landfills (e.g., paper).

- Identify options for managing the targeted portion of the waste stream (e.g., recycling).

- Estimate costs and benefits of diverting the waste.

- Estimate the amount of waste diverted over time and the avoided methane emissions from the avoided landfilling of the waste.

14.6 CONSTRUCTING BASELINE AND MITIGATION SCENARIOS

The purpose of this step is to estimate emissions with and without the implementation of selected mitigation options. It is recommended that these scenarios be developed in detail for the target subgroups for specific years, such as 2000, 2010, and 2025. The same models or techniques should be used to estimate both the baseline and the mitigation scenarios for each of the sources.

The inputs needed to develop the scenarios are discussed above. The underlying forecast variables used throughout the analysis should be used as the basis for the future estimates. Ranges of assumptions should be developed for those parameters or estimates that are uncertain. The implications of the uncertainty should be propagated throughout the assessment.

14.7 MITIGATION POLICIES

Both landfills and wastewater treatment facilities often fall within the purview of local, regional, or national governments. Waste disposal and treatment services are often provided directly by government agencies or are regulated activities of private entities. Consequently, implementation of emissions-reduction options for these sources will likely involve direct action on the part of governments.

14.7.1 Landfills

Landfill gas recovery and use can be implemented directly by government agencies that own and operate landfills. The technologies are well-known and the energy derived from the gas can be used to produce electricity or for other purposes. In cases where landfills are owned privately, options for promoting gas collection and use may include requiring gas collection and use by regulation and providing incentives for gas recovery and use, e.g., in the form of subsidies or guaranteed energy-purchase contracts.

Programs to divert waste from landfills would similarly be implemented by government agencies. Examples of approaches include the following:

- Increase landfill disposal costs for certain waste streams that can be diverted.
- Require paper products to contain a minimum percentage of recycled post-consumer fiber.
- Implement waste stream separation, either by waste generators or post-collection.
- Set source reduction and recycling targets for communities to attain.

14.7.2 Wastewater Treatment

Despite a widespread awareness of the health risks caused by poorly managed waste streams, many countries lack the resources with which to build a wastewater treatment infrastructure. Generally, wastewater treatment activities fall under the responsibility of local, regional, or national governments. Consequently, government initiatives to provide or improve wastewater treatment practices are likely to be called for. The broad government options for promoting the reduction of this source of methane are as follows:

- **Assist in, Provide for, or Promote the Development of Comprehensive Wastewater Management Policies, Infrastructure, and Treatment Systems.** In areas where no wastewater treatment systems exist, assistance could include policy development, funding and technological aid for the development of municipal collection and drainage systems, construction of municipal and industrial wastewater and sludge treatment facilities, and operation training programs.

- **Assist in the Design and Development of Smaller-scale, Community Wastewater Management Systems.** Small communities, where domestic wastes are often washed into streams or allowed to collect in gutters, latrines, or ponds, may account for a large part of methane emissions from developing countries. While complex treatment systems may not be feasible in such areas, smaller scale projects designed to divert waste streams into designated ponds and maintain aerobic or facultative (aerobic in the upper layers) conditions could reduce both methane emissions and health risks. Additional benefits of such projects may include reduced odors and the potential for using stabilized sludge as fertilizer.

- **Expand Management Infrastructure to Serve Entire Population.** Some regions and population segments may not be served by the existing waste management infrastructure. The expansion of treatment systems to these regions could potentially result in decreased methane emissions, as well as providing the benefits of improved wastewater management.

REFERENCES

Loehr, Raymond C. (1984), *Pollution Control for Agriculture,* second edition, Orlando: Academic Press, Inc. (Harcourt Brace Jovanovich, Publishers).

Richards, K.M. (1989), "Landfill Gas: Working with Gaia" from *Biodeterioration Extracts* no. 4, Energy Technology Support Unit, Harwell Laboratory, Oxfordshire, UK. December, 1989.

USAID (United States Agency for International Development) (1988), "Prospects in Developing Countries for Energy from Urban Solid Wastes," a *Bioenergy Systems Report,* September 1988, Office of Energy.

USEPA (United States Environmental Protection Agency) (1993a), *Options for Reducing Methane Emissions Internationally. Volume I: Technological Options for Reducing Methane Emissions,* U.S. Environmental Protection Agency, Office of Air and Radiation, K. Hogan (ed.), Washington, D.C.

USEPA (1993b), *Options for Reducing Methane Emissions Internationally. Volume II: International Opportunities for Reducing Methane Emissions,* U.S. Environmental Protection Agency, Office of Air and Radiation, K. Hogan (ed.), Washington, D.C.

USEPA (1993c), *Opportunities to Reduce Anthropogenic Methane Emissions in the U.S.,* U.S. Environmental Protection Agency, Office of Air and Radiation, K. Hogan (ed.), Washington, D.C.

USEPA (1994), *International Anthropogenic Methane Emissions: Estimates for 1990,* Report to Congress, U.S. Environmental Protection Agency, Office of Policy, Planning and Evaluation, M.J. Adler (ed.), Washington, D.C.

Vogler, Jon A. (1984), "Waste Recycling in Developing Countries," in *Managing Solid Wastes in Developing Countries,* John Holmes, (ed.), John Wiley and Sons, Ltd.

CHAPTER 15
REPORTING A MITIGATION ASSESSMENT

15.1 INTRODUCTION

The results of a mitigation assessment may be presented in various ways depending on the nature of the assessment. It is suggested that the results be presented separately for the energy and non-energy sectors. If an integration of energy and non-energy analyses has been conducted, these results should also be presented and discussed. Finally, a summary should present the main findings from each sector as well as from integrated analyses.

15.2 ENERGY SECTOR

15.2.1 Introduction

The following suggestions for presenting a mitigation assessment for the energy sector are fairly extensive and should be looked upon as general guidelines. In some countries, lack of data and limitations in model structure may make it difficult to include the level of detail suggested here; presentation should be tailored to country-specific conditions.

It is important that the presentations not only focus on the results of the assessment but also discuss significant assumptions made when modeling the energy system. It should also present background information and data used in the analysis.

15.2.2 Model Description

Describe the model and methodology used for the mitigation assessment. Describe how the structure of the national energy system is constructed using the modeling tool(s).

Discuss the sectoral breakdown that is used in the study and the level of disaggregation within sectors.

15.2.3 Scenario Assumptions and Input Data

15.2.3.1 General scenario assumptions

Macro-parameters. Discuss the overall macro-parameters—GDP and population and their distribution—that drive energy use. Present projections for these over the whole analysis period.

Price Assumptions on Imported and Exported Fuels. Present assumed price paths for imported fuels, such as crude oil, oil products, and natural gas. If applicable, present prices assumed for exported energy carriers.

Discount rate(s). Present the discount rate(s) used in the analysis. If different discount rates are used in different sectors (to reflect different market barriers, capital scarcity, etc.), discuss how the values for these were established.

15.2.3.2 Projections of activity levels

Present the base year and future development of activity levels (e.g., value added, residential area, passenger-km, tonne-km, etc.) for each sub-sector within the main sectors of transport, industry, residential, and services.

Describe how the projections were established. Explain how the macro-parameters drive the activity levels in the different demand sectors and sub-sectors. Discuss important assumptions like development of infrastructure for transportation. If a macro-economic model is used for projecting the activity levels, give a short description of, and the assumptions used in, the model.

15.2.3.3 Projection of energy intensities

List assumptions and sources used when estimating the base year intensities and assumptions made for projecting future development. For simulation and optimization models, where the choice of technologies is made within the model, it is important to explain to what degree improvements in technology efficiencies are included in the projected intensities as opposed to given as specific technology options. Similarly, in accounting models, it is important to distinguish between improvements included in the baseline intensities and improvements included in the mitigation options.

Describe (if applicable) the historical time-series used to project future energy intensities. If the assumed development of the projected intensities differs between the baseline and various mitigation scenarios, explain what these differences are assumed to reflect.

15.2.3.4 Description of energy resources and technologies

Discuss the process of screening technology options:

- **Energy Resources**. Describe the resources for primary energy carriers included in the analysis. Present future production profiles for domestic extraction of oil, gas, and coal, if applicable. Discuss energy use and GHG emissions resulting from the production and possible options for mitigation in the production processes, e.g., reduced flaring of associated gas.

  Present price assumptions for domestic fuels and discuss limitations on availability, (e.g., due to pipeline capacity).

  For each source of biomass, present estimates of resource availability that can be utilized for energy purposes. Describe the resource potential as a function of the cost of producing and collecting biomass. Discuss environmental impacts of the extraction and competing non-energy use of the biomass.

  Describe other renewable resources, like wind, hydro, and solar, similarly.

- **Energy Supply Technologies**. Present data describing the main features of each of the options. These include data on investment costs, variable and fixed O&M costs, efficiency, emission coefficients, maximum market penetration rates, if any, and installed capacity in base year. Specify data that are time-dependent, e.g., reduced costs for PV-cells over time or improved efficiency in electricity grid.

- **Energy Demand Technologies.** Present data describing the main features of the technology options selected for each end-use sector. This includes data for investment costs, variable and fixed O&M costs, efficiency, emission coefficients, maximum market penetration rates, if any, and installed capacity in base year.

15.2.3.5 Emission coefficients

Present the emission coefficients for technologies and/or energy carriers used in the study. If GHGs other than CO_2 are considered, describe the assumptions that are made about global warming potential for each gas, expressed in CO_2-equivalents. If emissions of non-GHGs are included in the study, list emission coefficients for the gases considered.

15.2.4 Scenario Definitions

15.2.4.1 Baseline scenario

Describe key assumptions in the scenario, including policy measures such as insulation standards and efficiency programs, expansion plans for electricity production, and taxation or subsidy of energy products.

If an optimization model is used for the analysis, discuss to what degree the baseline is constructed exogenously as opposed to allowing the model to shape the baseline through choosing among the same options as are available in the mitigation scenarios.

15.2.4.2 Mitigation scenarios

Describe how the assumptions for the mitigation scenarios differ from the baseline. Examples of types of scenarios include:

- GHG-constraints (e.g., 20% reduction from baseline by 2010);
- GHG-"taxes" (by making the model add a cost per unit of emission);
- focus on specific technologies (nuclear, renewable, etc.); and
- other assumptions that differ from the baseline (e.g., subsidies for targeted technologies).

15.2.5 Results

15.2.5.1 GHG emissions

Present emission levels for the baseline and all mitigation scenarios. If GHGs other than CO_2 are considered, show emissions by type of gas expressed in CO_2-equivalents.

When a dynamic model is used, the emissions should be presented as a function of time for the whole time horizon. If a "snap-shot" model is used, present base year emissions and the target years studied. Show breakdown of the emissions into main sectors for the baseline and the mitigation scenarios.

All scenarios should have the same emissions in the base year. Explain any deviations from the emission inventories in this year.

15.2.5.2 Energy use

Primary Energy. Present the primary energy requirements[1] for the different scenarios for all years studied. Show the shares of primary energy sources (oil, natural gas, coal, nuclear, biomass, hydro, other renewables). Discuss implications for imports and exports and the depletion of domestic resources.

Electricity Generation. Discuss the development of capacity expansion of the power system and the types of technologies being used in the various scenarios. Address improvements in transmission and distribution, and peaking requirements.

Present the figures for annual electricity production and the fuel mix in the power production. Discuss the estimates of capacity utilization.

Show figures for import/export of electricity, if applicable.

Final Energy. The final energy use should be presented for each scenario and for each main sector.[2] Show breakdown into major fuels and electricity. Describe differences between the scenarios and discuss impacts of end-use efficiency.

Final energy intensities should be presented in terms of final energy use per unit of activity level for each sector (e.g., PJ/$ value added, PJ/person-km, PJ/household, etc.). Also present intensities for important end uses.

15.2.5.3 Cost of emission abatement

The presentation of the costs of emission abatement will to some degree depend on the model that is used in the analysis. For the pure engineering models that do not include feedback to the general economy, the costs only include technical costs of the energy system, as opposed to the hybrid model MARKAL-MACRO and the general equilibrium model LBL-CGE that also include impacts on GDP. The cost representation in the engineering models can also vary. For example, the optimization models ETO and MARKAL have the option to select technologies on the basis of the total cost of industrial production rather than on the basis of energy costs alone.

---

[1] Explain how the primary energy requirements are calculated. Which conversion factors are used for electricity production from hydro, solar, wind power, and nuclear power?

[2] Final energy use refers to actual consumption by end users. This includes purchased fuels and electricity, consumption of biomass, and uses of solar and wind energy. Final energy excludes all transmission and distribution losses.

The presentation of costs should include a short description of what elements the cost figures include:

Additional Energy System Costs. Present the differences in total system cost between each mitigation scenario and the baseline. If a multiperiod, dynamic model is used, present the cost increase as a function of time, as well as the cumulative costs for the whole time horizon considered. For putting the costs in perspective, the cost figures can be expressed as percentages of GDP in the specific years.

Cost Structure. The total cost of the energy system can be divided into three components:

- investment costs to replace and expand the capital stock in energy supply, conversion, and end-use sectors;
- fuel supply costs, broken into expenditures for domestic fuels and for (net) imported fuels; and
- other costs, like operating and maintenance costs.

In addition to understanding how the total costs change when moving to lower emission levels, it is important to monitor how the cost components change. Typically the capital cost is the most important component when costs increase in mitigation scenarios. Nuclear energy, renewable technologies, more efficient equipment, and investment in energy savings all drive up investment costs as capital substitutes for fossil fuel. In some cases, as when using expensive carbon-free hydrogen instead of fossil fuels, the fuel costs can also increase.

The increase in investment costs can be crucial in developing countries where capital is scarce. For example, if the total system costs increase by 0.5% of GDP in a mitigation scenario, and the increase in costs consists of 1% higher capital costs and 0.5% lower fuel costs, the extra capital costs may raise the total investments in the economy (typically around 15-30% of GDP in developing countries) by as much as 4-7%.

To the extent information about implications on foreign exchange for different technology options is included in the model, the total impacts on foreign exchange should be presented for each scenario. Even if the information on foreign exchange is not directly available from the model, the impacts on the foreign exchange from fuel imports and exports can be reported by presenting the net expenditure on imported fuels.

Cost Curves. Cost curves for emission abatement express the costs per unit of emission reduction as a function of the quantity of GHG reduced. As discussed in Chapters 2 and 3, these curves can be established in different ways, depending on which model is used and the level of detail in the study. We recommend that the integrated approach is used to the extent possible.

A cost curve for average cost, reflecting the cost difference between a mitigation scenario and baseline, should be presented. When optimization models are used, the marginal cost curve should also be presented. For the other models, the incremental cost curve can be used as an approximation of the marginal cost curve. (See Chapter 2 for definitions of average, incremental, and marginal costs.) Refer to figures in Chapter 2 and 3 for examples of cost curves.

The costs of GHG abatement, and thus the cost curves, will typically vary significantly between different time periods. Cost curves should therefore, when possible, be presented for more than one year or time period. When using a dynamic model, the cost curves can also represent cumulative reductions over the entire time horizon studied.

15.2.5.4 Contribution of technology options to GHG abatement

Energy Supply Options. For each scenario, present the mix of technologies used in generation of electricity and technologies used in production, transportation, and processing of energy carriers (e.g., enhanced oil recovery techniques, refinery processes, gas pipelines, charcoal production, etc.).

End-use Options. For each end-use sector, present the mix of technology options contributing to satisfying the demand for energy. Focus on the differences between scenarios, and explain major fuel shifts and impacts of conservation and higher equipment efficiencies.

15.2.5.5 Other environmental impacts

If included in the study, present emissions of other pollutants such as SO_x and NO_x.

Discuss impacts on land-use in the various scenarios (important with hydropower).

15.2.5.6 Summarize evaluation of options

Summarize the technology options that are attractive for emission abatement under different scenario assumptions. If criteria other than cost are considered, describe the characteristics of the options with respect to these criteria. If a multiperiod model is used, address in which time periods the various options are most attractive. Discuss timing of investment in new capacity.

Assess the robustness of the options chosen in the mitigation scenarios and the uncertainty of the data used.

15.2.6 Macroeconomic Impacts

Describe the type of macroeconomic analysis conducted (if any).

Discuss feedback to the general economy from the results of the energy sector analysis. This can be done through checking against assumptions on macro-parameters and development of activity levels. If applicable, iterate through modifying key parameters.

Discuss capital requirements and impacts on foreign exchange in the different mitigation scenarios.

15.2.7 Policy Options

Discuss ways to implement the most attractive and promising mitigation options and the possible barriers for implementation. Distinguish between short-term and long-term considerations.

Describe the type of evaluation of policies conducted (if any).

15.3 NON-ENERGY SECTORS

15.3.1 Introduction and Land-Use Framework

A description of methods and results should be presented for each sector studied. A report should first present the overall land-use context as it affects the forest, agriculture, and rangelands/grasslands sectors. This section is critical since it sets the basis for the analysis of the non-energy sectors.

Discuss the base-year allocation of land for various uses, including national forests, urban areas, agriculture, pasture, hydro reservoirs, etc. Present and discuss:

- base-year situation and future potential for the intensity with which different types of land may be used for various purposes;
- projected demand, for one or more target-years, for land-based products such as cattle, food products, mineral extraction, and urban use; and
- the future land use pattern in baseline and mitigation scenarios, and, if applicable, the need to import products should the land area be inadequate to meet domestic needs.

15.3.2 Sectoral Mitigation Assessment

The presentation format described below can generally be used for each of the non-energy sectors. The text below refers to carbon sequestration and is oriented toward the forest sector, but a similar presentation can be made for assessment of methane abatement and carbon sequestration options in other sectors.

- **Model/Method Description**. Describe the model(s) or method(s) used for the assessment.
- **Scenario Assumptions and Input Data**. Describe the key assumptions used to develop a baseline scenario and the nature of the input data.
- **Mitigation Options**. Describe and discuss the mitigation options that were selected for analysis. Explain the reasons for their selection. For the forest sector, explain whether a sustainable rotations or plant and store approach was used to evaluate the impacts of mitigation options.
- **Carbon Sequestration**. Describe and discuss the analysis of carbon flows for each mitigation option. Explain the assumptions regarding carbon content of biomass, biomass density, soil carbon, survival rates, rotation period, etc., which were used to arrive at carbon flow estimates for vegetation, soil, forest product, and decomposing matter. Present both base year and future time period estimates for the baseline and mitigation scenarios.
- **Monetary Costs and Benefits**. Describe and discuss the analysis of monetary flows for each mitigation option. Explain the assumptions regarding unit costs, product prices, discount rates, and opportunity cost of land. Present both base year and future time period estimates.

  Present estimates of the costs and benefits of the mitigation options for different participants.

- **Comparison of Monetary and Carbon Flows.** Present indicators of cost-effectiveness which compare monetary and carbon flows. These include the initial cost, establishment cost, and net present value of benefit per unit of carbon protected or sequestered.

  Present cost of conserved carbon curves, which show the cost-effectiveness indicator plotted against the amount of carbon sequestered or protected for a series of mitigation options.

- **Scenarios of Carbon and Monetary Flows.** Present scenarios of future carbon and monetary flows until the final year chosen for analysis. Present estimates of both cumulative and annual flows.

- **Macroeconomic Implications.** Discuss the key macro-economic issues such as GDP, value of timber- and forest-products, international trade, and jobs creation.

- **Barriers to Implementation and Policy Options.** Discuss the main barriers to implementation and the national and local policy interventions that may be required to implement the mitigation options. The discussion should focus on economic, regulatory, and infrastructure policies.

15.4 SUMMARY OF ENERGY AND NON-ENERGY MITIGATION ASSESSMENTS

The presentation of a national mitigation assessment should contain a section that summarizes the results from the analyses done for the energy and non-energy sectors. A summary should present the key results from each of the sectoral assessments, particularly the GHG impacts, costs, and other impacts of specific mitigation options or groups of options. The total additional cost of the mitigation scenarios should be reported for each sector, and the evolution of costs over time should be graphed if such data are available. A ranking of options in each sector, based on both quantitative and qualitative criteria, may also be presented. Information on other environmental impacts and indirect economic impacts should be reported for each scenario or for important mitigation options. Results from top-down analysis of mitigation costs should also be presented if such analysis has been conducted. The summary should also discuss policies and programs that could encourage adoption of mitigation options in each sector.

Analyzing the results of assessments for the energy and non-energy sectors in an integrated framework is a challenging task since the methods and many of the assumptions will differ among the sectoral studies. In reporting the results of such an integrated analysis, the approach used should be carefully explained. Whenever impacts of GHGs other than CO_2 are presented in terms of CO_2-equivalent, the factors used to convert to GWP values should be clearly stated.

15.4.1 Summary Description of Mitigation Options

The costs, GHG impacts, and other effects of specific mitigation options should be reported for the most attractive options in each sector. This reporting will permit a comparison of the various options in terms of the criteria that are of most importance. Table 15-1 illustrates a sample format. Costs should be reported in terms of capital investment, life-cycle cost, and net cost (net of monetizable benefits). In practice, many of the values will be presented as a range.

Estimates of unit cost (cost/tCeq) should be reported for each mitigation option. In the energy sector in particular, it can be difficult to estimate a specific cost per tonne of carbon-equivalent since the cost of any given option is dependent on the configuration of the total system (see discussion of cost curves in Chapters 2 and 3). In this case, a range of costs should be given for each option.

15.4.2 National GHG Emissions

Summing results across sectors can yield estimates of total gross and net (after subtracting carbon sequestration) GHG emissions in a baseline scenario. Selected options from different sectors can also be grouped into national-level mitigation scenarios, although ensuring consistency of assumptions across sectors can be problematic. For example, a cross-sectoral mitigation scenario could include all options up to a given cost per tonne of carbon-equivalent. As discussed in Chapter 2, cost curves from different sectors can be combined in order to identify the options that provide a least-cost solution for satisfying a specific emission reduction target.

15.5 CONCLUSION

The reporting of a mitigation assessment should be guided by consideration of the audiences for whom the report will be of interest. One audience consists of professional colleagues in the scientific community both within a country and in other countries. In many cases, a mitigation assessment will become a definitive analysis for the country to which it applies and will serve as a reference for the international community of scientists involved in work related to global climate change. Thus, the reporting of an assessment should have a high level of scientific rigor and good documentation of methods and data sources; the degree of uncertainty associated with the findings should also be discussed.

Perhaps the most important audience for a mitigation assessment consists of policy-makers who may take various actions based on the results of the assessment. For this audience, the comprehensibility of the report and the clarity of its conclusions is of utmost significance. The results need to be presented in a way that is relevant to policy-makers. In addition, a report must strike a balance between acknowledging the uncertainty associated with the findings and drawing conclusions that can form a basis for policies and programs.

Table 15-1: Sample Format for Reporting the Impacts of Mitigation Options

Criteria	Mitigation Option 1	Mitigation Option 2	Mitigation Option n
GHG saving or storage (tCeq) – Carbon (tC) – Methane (tCeq) – Other gases (tCeq)	100,000	150,000	20,000
Life-cycle-cost (currency/tCeq) Capital investment (currency/tCeq) Net costs (currency/tCeq)	3.5 2.0 3.3	40 30 40	120 80 110
Indirect economic impacts – Jobs created (#) – Reduced imports (US$)	300,000	100,000	Uncertain
Equity considerations, impact on – Low-income jobs – Low-income monthly expenditure	Low High	Medium Low	Uncertain None
National environmental impacts (net change) – Sulfur oxides (tonnes) – Particulates (tonnes) – Biodiversity	-150 -30 No effect	-200 None Reduction	-75 -50 Uncertain
Potential impact of implementation policies	Low	Medium	High
Sustainability of option	High	Uncertain	Medium
Consistency with national development goals	High	Low	Medium
Uncertainty of data – Technology performance and costs – Costs of implementation programs	Low High	Medium Low	High Medium

GLOSSARY

Afforestation - Planting of new forests on lands which historically have not contained forests. These newly created forests are included in the category "Managed Forests" in the Land-use Change and Forestry module of the emissions inventory calculations. See also reforestation.

Albedo - The surface reflectivity of the globe. Affects the amount of solar radiation being radiated back into space without being absorbed by the earth's climate system.

Anthropogenic emissions - Emissions resulting from human activities.

Appliance - Any household energy-using device.

Base year - The year for which the inventory is to be taken. This is currently 1990. In some cases (such as estimating CH_4 from rich production), the base year is simply the last year of a number of years over which an average must be taken.

Biodiversity - Biological diversity, i.e., the variety of species in a given area.

Biofuels - Wood, waste, and alcohol fuels.

Biomass - Organic material both above the ground and below ground, and both living and dead, e.g., trees, crops, grasses, tree litter, roots, etc. When burned for energy purposes, these are referred to as biomass fuels.

Biosphere - Refers to the zone of the earth and atmosphere that contains living organisms. The terrestrial biosphere excludes the oceans.

Bottom-up modeling - A modeling approach which arrives at economic conclusions from an analysis of the effect of changes in specific parameters on narrow parts of the total system.

Carbon tax - A tax on fossil fuels based on the individual carbon content of each fuel. Under a carbon tax, coal would be taxed the highest per MBtu, followed by petroleum and then natural gas.

Carbon cycle - General term used in reference to the sum of all reservoirs and flows of carbon on Earth. The flows tend to be cyclic in nature; for example, carbon removed from the atmosphere (one reservoir) and converted into plant tissue (another reservoir) is returned back into the atmosphere when the plant is burned.

Carbon reservoir or sink - Within the carbon cycle, the physical site at which carbon is stored (e.g., atmosphere, oceans, Earth's vegetation and soils, and fossil fuel deposits).

CFCs (Chlorofluorocarbons) - A family of inert gases, including CFC-11, CFC-12, and CFC-113.

Climate - The statistical collection and representation of the weather conditions for a specified area during a specified time interval (usually decades).

Closed forest - A dense forest with closed canopy through which sunlight does not penetrate sufficiently for grasses to grow on the forest floor. These forests contain a significantly greater amount of biomass per hectare than open forests.

Cogeneration - The simultaneous generation of both electric power and heat; the heat, instead of being discharged without further use, is used in some fashion (e.g., in district heating systems).

Cultivar - Variety of plant species.

Deforestation - Converting forest land to other vegetation or uses (e.g., cropland, pasture, dams).

Degradable organic carbon - Organic material which can decay, expressed as weight of carbon. Usually 15 to 25% of total waste.

Demand-side management - The planning, implementation, and monitoring of utility activities designed to encourage customers to modify their pattern of electricity usage.

Discount rate - The rate at which money grows in value (relative to inflation) if it is invested.

Dynamic - In the field of modeling, a dynamic model includes inter-temporal relations between variables. A model that does not include such relations is called static.

Dynamic programming - A method to find an optimal time path.

Emission factor - A coefficient that relates actual emissions to activity data as a standard rate of emission per unit of activity. Emission factors are often based on a sample of measurement data averaged to develop a representative rate of emission for a given activity level under a given set of operating conditions.

Endogenous variables - Variables determined within the system under consideration.

Energy forms and levels - Primary energy is energy that has not been subjected to any conversion or transformation process. Secondary energy (derived energy) has been produced by the conversion or transformation of primary energy or of another secondary form of energy. Final energy (energy supplied) is the energy made available to the consumer before its final conversion (i.e., before utilization). Useful energy is the energy made usefully available to the consumer after its final conversion (i.e., in its final utilization).

Energy intensity - The amount of energy required per unit of a particular product or activity.

Energy services - The service or end use ultimately provided by energy. For example, in a home with an electric heat pump, the service provided by electricity is not to drive the heat pump's electric motor but rather to provide comfortable conditions inside the house.

Engineering approach - A particular form of bottom-up modeling in which engineering-type process descriptions (e.g., fuel efficiency of end-use devices) are used to calculate a more aggregated energy demand. This term is particularly used in contrast to econometric models.

Enteric fermentation - The intestinal fermentation which occurs in ruminant animals such as cows; it is a major biological source of methane.

Exogenous variables - Variables which are determined outside the system under consideration. In the case of energy planning models, these may be political, social, environmental, and so on.

Feedback - When one variable in a system (e.g., increasing temperature) triggers changes in a second variable (e.g., cloud cover) which in turn ultimately affects the original variable (i.e., augmenting or diminishing the warming). A positive feedback intensifies the effect. A negative feedback reduces the effect.

Fossil fuel - Coal, petroleum, or natural gas or any fuel derived from them.

General equilibrium analysis - An approach which considers simultaneously all the markets in an economy, allowing for feedback effects between individual markets. It is particularly concerned with the conditions which permit simultaneous equilibrium in all markets, and with the determinants and properties of such an economy-wide set of equilibrium.

Greenhouse effect - An atmospheric process by which greenhouse gases (such as CO_2, CH_4, N_20, and CFCs) affect the global energy balance. Shortwave radiation from the sun that reaches the earth and is re-emitted as longwave infrared radiation is partially absorbed by greenhouse gases (GHGs). In the absence of GHGs, the earth's average temperature would be 18° C rather than 15° C .

Greenhouse gas - Any gas that absorbs infrared radiation in the atmosphere.

Global Warming Potential (GWP) - Some greenhouse gases are more effective, on a unit basis, of affecting, or "forcing," the climate system. The GWP combines the capacity of a gas to absorb infrared radiation and its residence time in the atmosphere with a time frame of analysis, then expresses the result relative to CO_2.

Income elasticity - The expected percentage change in the quantity demand for a good given a one percent change in income. An income elasticity of demand for electricity of 1.0 implies that a one percent increase in income will result in a one percent increase in demand for electricity.

Input-output analysis - Method of investigating the interrelationship between the branches of a national economy in a specific time period. The representation, in the form of a matrix table, is called an input-output table. An input-output analysis allows the changes in total demand in related industrial branches to be estimated.

Least-cost planning - In energy planning, the practice of basing investment decisions on the least-costly option for providing energy services. It is distinguished from the more traditional approach taken by utilities, which focuses on the least-costly way to provide specific types of energy, with little or no consideration of less-costly alternatives that provide the same energy service at lower costs.

Life-cycle cost - The cost of a good or service over its entire life cycle.

Linear programming - A practical technique for finding the arrangement of activities which maximizes or minimizes a defined criterion subject to the operative constraints. For example, it can be used to find the most profitable set of outputs that can be produced from a given type of crude oil input to a given refinery with given output prices. The technique can deal only with situations where activities can be expressed in the form of linear equalities or inequalities, and where the criterion is also linear.

Macro-economics - The study of economic aggregates and the relationships between them. The targets of macro-economic policy are the level and rate of change of national income (i.e., economic growth), the level of unemployment, and the rate of inflation. In macro-economics, the questions about energy are how its price and availability affect economic growth, unemployment, and inflation; and how economic growth affects the demand for energy.

Manure - Waste materials produced by animals that are managed for agricultural purposes. When manure is managed in a way that involves anaerobic decomposition, significant emissions of methane can result.

Marginal costs - In a linear programming environment, this term has the very specific meaning of change of the objective function value as a result of a change in the right-hand-side value of a constraint. If, for example, the objective is to minimize costs, and if the capacity of a particular energy conversion facility, such as a power plant, is fully utilized, the marginal cost in the linear programming sense expresses the (hypothetical) reduction of the objective function value (i.e., the benefit) of an additional unit of capacity.

Market clearing - The economic condition of supply equaling demand.

Open forests - Open forests are less dense than closed forests, do not have a closed canopy, and have grasses growing on the forest floor. These forests contain less biomass per hectare than closed forests.

Optimization model - A model describing a system or problem in such a way that the application of rigorous analytical procedures to the representation results in the best solution for a given variable(s) within the constraints of all relevant limitations.

Price elasticities - The expected percentage change in quantity demand for a good given a one percent change in price. A price elasticity of demand for electricity of -0.5 implies that a one percent increase in price will result in a half percent decrease in demand for electricity.

Radiative forcing - Changes in the global balance of incoming solar radiation and outgoing infrared radiation caused by a radiative forcing agent, such as clouds, surface albedo, and greenhouse gases. This results in changes in the global climate.

Reforestation - Planting of forests on lands which have, historically, previously contained forests but which have been converted to some other use. Replanted forests are included in the category "Managed Forests" in the Lands Use Change and Forestry module of the emissions inventory calculations. See also afforestation.

Renewable energy - Energy obtained from sources that are essentially inexhaustible (unlike, for example, the fossil fuels, of which there is a finite supply). Renewable sources of energy include wood, waste, wind, geothermal, and solar thermal energy.

Retrofit - To update an existing structure or technology by modifying it, as opposed to creating something entirely new from scratch. For example, an old house can be retrofitted with advanced windows to slow the flow of energy into or from the house.

Ruminant animals - Herbivores (grazing animals such as cattle, buffalo, sheep, goats, and camels) which have a large free stomach or rumen. Digestion in anaerobic conditions in the rumen can create significant emissions of methane from ruminant animals.

Scenario - Coherent and plausible combination of hypotheses, systematically combined, concerning the exogenous variables of a forecast.

Sensitivity analysis - A method of analysis which introduces variations into a model's explanatory variables in order to examine their effects on the explained.

Sequester - To isolate and remove something. As used here, the processes by which carbon dioxide is removed from the atmosphere and retained for some period in a carbon reservoir (e.g., trees).

Simulation model - Descriptive model based an a logical representation of a system and aimed at reproducing a simplified operation of this system. A simulation model is referred to as static if it represents the operation of the system in a single time period; it is referred to as dynamic if the output of the current period is affected by evolution or expansion compared with previous periods. The importance of these models derives from the impossibility or excessive cost of conducting experiments on the system itself.

Sustainable - A term used to characterize human activities that can be undertaken in such a manner as to not adversely affect the environmental conditions (e.g., soil, water quality, climate) necessary to support those same activities in the future.

Temperate - Relating to the region between the tropics and the polar circles (between 23.5° and 66.5°) in both hemispheres.

Top-down modeling - A modeling approach that proceeds from broad, highly aggregated generalizations to regionally and/or functional disaggregated details.

Tropical - Relating to the region between the Tropic of Cancer and the Tropic of Capricorn, 23.5° North and 23.5° South, respectively.

User interface - All information, including push-button help texts, facilitating the technically correct operation of a computer program.

GPSR Compliance

The European Union's (EU) General Product Safety Regulation (GPSR) is a set of rules that requires consumer products to be safe and our obligations to ensure this.

If you have any concerns about our products, you can contact us on ProductSafety@springernature.com

In case Publisher is established outside the EU, the EU authorized representative is:

Springer Nature Customer Service Center GmbH
Europaplatz 3
69115 Heidelberg, Germany

Batch number: 10824712

Printed by Printforce, the Netherlands